SELLING

SCIENCE

SELLING SCIENCE

HOW THE PRESS COVERS SCIENCE AND TECHNOLOGY

DOROTHY NELKIN

W. H. FREEMAN AND COMPANY

NEW YORK

Q
225
.N35
1987

Library of Congress Cataloging-in-Publication Data

Nelkin, Dorothy.
    Selling science.

    Includes index.
    1. Science news.   2. Communication in science.
3. Communication of technical information.   I. Title.
II. Title: Press covers science and technology.
Q225.N35   1987   070.4'495   86-25820
ISBN 0-7167-1826-X

Printed in the United States of America

1 2 3 4 5 6 7 8 9 0   MP   5 4 3 2 1 0 8 9 8 7

C O N T E N T S

PREFACE   xi

1

IMAGES OF SCIENCE AND TECHNOLOGY   1

2

THE MYSTIQUE OF SCIENCE IN THE PRESS   14

The Scientist as Star   15
Science as a Resource   21
The Purity of Science   24
The Authority of Scientific Theory   27

3

THE PRESS ON THE TECHNOLOGICAL FRONTIER   33

Advertising High Technology   34
Problems and Promises of the Technological Fix   41

4

THE PERILS OF PROGRESS   53

The Ozone Controversy   55
The Sweetener Dispute   58
The Dioxin Debate   64

5

MEDIA MESSAGES, MEDIA EFFECTS   70

Media Influence on Public Attitudes   72
The Framing of Public Policy   80

6

THE CULTURE OF SCIENCE JOURNALISM   85

The Development of a Style   86
Norms of Objectivity   91
Changing Professional Ideals   96
Social Biases of Science Writers   100

7

CONSTRAINTS OF THE JOURNALISTIC TRADE   109

Newswork   111
Editorial Constraints   114
Audience Assumptions   118
Economic Pressures   120
Constraints of Complexity   124
Vulnerability to Sources   128

8

THE PUBLIC RELATIONS OF SCIENCE 132

Promoting Scientific Institutions 134
Scientists in Industrial Public Relations 144

9

HOW SCIENTISTS CONTROL THE NEWS 154

Blaming the Messenger 155
Strategies of Control 159

10

THE HIGH COST OF HYPE 170

NOTES 183

INDEX 213

P R E F A C E

This book reflects my conviction that fair, critical, and comprehensive reporting about science and technology is extremely important in a society increasingly dependent on technological expertise. The importance of public understanding of science has been widely recognized in the growing number of discussions of science literacy and the persistent complaints about the problems of science education. However, most of these discussions focus on science education in the schools. Although most adults in fact become informed about science and technology through the media, there has been little analysis of the way science is portrayed by journalists or of the relationship between the two influential social institutions of science and the press.

My own interest in this topic developed while I was doing research on public attitudes toward science and technology and, in particular, on technological controversies. I was struck by the ubiquitous tendency to blame the press: scientists, engineers, and physicians are quick to condemn the media, to criticize the quality of science reporting, and to attribute negative or naive public attitudes toward science and technology to the images conveyed in the press. Yet, at the same time, they are often unable to document their complaints, to specify what is wrong. I therefore began to explore the images of science and technology that are conveyed to the public through the press, and the characteristics of both journalism and science that contribute to shaping these images.

My analysis concentrates on the popular press, including (1) national newspapers such as the *New York Times,* the *Christian Science Monitor,* the *Wall Street Journal,* and the *Washington Post;* (2) local newspapers, drawing from *Newsbank,* a file of 100 local newspapers around the United States that is indexed by categories to cover local views of selected public issues; (3) national news magazines such as *Time, Newsweek,* and *U.S. News and World Report;* and (4) widely distributed specialized magazines, including women's, health, and business magazines. I have excluded from my analysis specialized science magazines, such as *Science* and *Scientific American,* which are directed to an audience already informed, engaged, and especially interested in technical subjects. I have included only limited material on television, drawing on secondary sources for comparative purposes. Although science fiction programs, medical dramas, docudramas, and many of the soaps present fictionalized images, relatively little information about specific developments in science and technology is conveyed by television. Educational programs such as "NOVA" popularize science, but they have a limited audience and are not a source of regular science news. People tend to go to newspapers and magazines to learn about current science and technology events.[1]

To analyze the coverage of science and technology in the popular press, I selected a spectrum of issues that have attracted extensive and widespread coverage over the past two decades. I looked for the dominant themes and recurring metaphors that set the tone of science journalism and project an image of science and technology to the reader. For the sake of brevity, many areas are neglected, the most striking being the military. The unique political character of this area of technology warrants a separate project, and such work is under way at New York University.[2]

The literature on the public communication of science and technology is scattered. This book is in part an effort to bring together the existing studies that bear on the complex relations between scientists and journalists as they influence the images of science in the press. But my investigation led well beyond this literature. For material on scientists' concern about their image in the press I turned to science policy journals and to the professional journals of science, engineering, and medicine. Some of the material on the public relations activities of scientists and industries was gathered directly from public relations firms; some from discussions with scientists, journalists, and public relations officers; some from participation in meetings on the subject. I also drew on information from my earlier studies of scientific and technological controversies.

To better understand the nature of science journalism I spoke extensively to many reporters and sat in on various meetings, press conferences, group discussions, and informal affairs. As a member of the Council for the Advancement of Science Writing, I have had the privilege of spending many hours with some of the leading American science journalists and of listening to their concerns about their profession. They have tolerated a flood of naive questions from me for several years. I also gathered material from the Media Resources Service of the Scientists' Institute for Public Information, which serves as a liaison between scientists and journalists. Those quotations in the text that are not specifically referenced are from informal discussions with journalists and scientists.

I benefited greatly from transcripts of interviews with journalists conducted by Sharon Friedman. I also used news files that were collected by colleagues who have done research on the topics covered in Chapters 2 through 4. These people include Jon Beckwith, Stephen Hilgartner, Gerry Markle, Mary Marlino, Elena Nightingale, and Chris Anne Raymond. Roald Hoffmann and Kenneth Wilson made available their clippings on their Nobel prizes. David Perlman and other journalists gave me access to their files. I also gained insights from the discussions of the Twentieth Century Fund Task Force on the Communication of Risk, for which I wrote the background paper (*Science in the Streets*, New York: Priority Press, 1984).

I am particularly indebted to Stephen Hilgartner for research assistance and for creative guidance in analyzing metaphors and images. Several people provided invaluable criticism of early drafts of this manuscript, including Barbara Culliton, Sharon Dunwoody, Daniel X. Freedman, Sharon Friedman, Rae Goodell, Fred Jerome, Alex Keynan, Marcel LaFollette, Elena Nightingale, David Perlman, Carol Rogers, David Rubin, Michael Schudson, and David Zimmerman. My editor, Jonathan Cobb, contributed far more than editing, providing ideas and a great deal of help in shaping the final draft of the book. Sandra Kisner patiently edited and typed many drafts.

Finally, research expenses and time for research and writing was provided by a Guggenheim Fellowship, a year as a visiting scholar at the Russell Sage Foundation, and a grant (R11-8510076) from the National Science Foundation.

Dorothy Nelkin
Ithaca, New York
September 1986

1

IMAGES OF SCIENCE
AND TECHNOLOGY

"True descendants of Prometheus, science writers take the fire from the scientific Olympus, the laboratories and the universities, and bring it down to the people."[1] This was how one of the earliest science journalists, William Laurence of the *New York Times*, described his profession. When Laurence began his career in the 1930s, Prometheus had few descendants; Laurence was one of only about a dozen science writers in the United States. And fire from scientific Olympus seldom appeared in the press.

Today the popular press is paying increased attention to science and technology. Eighteen daily newspapers with a total circulation of seven million have added special weekly science sections in the past decade, sixteen having appeared since 1982. In 1956 William Laurence was one of three science journalists on the *New York Times*.[2] Today there are twelve. Science and technology are discussed in feature articles as well as in news about economic and business affairs. The growing policy importance of science-related issues—environmental quality and public health, regulation of food additives and drugs, the siting of large-scale technical and research facilities, the impact of new medical techniques—is reflected in regular, sometimes extensive press coverage. Technology-related crises—Bhopal, Three Mile Island, Chernobyl, Love Canal, the explosion of the space shuttle—are frequently front-page news.

Public understanding of science and technology is critical in a society increasingly affected by their impacts and by policy decisions determined by technical expertise. At the community level, people are continually confronted with choices that require some understanding of scientific evidence: whether to allow the construction of a nuclear plant or a toxic waste disposal dump, whether to tolerate a child with AIDS in their schools. Similar choices must be made at the personal level: whether to use the pill, whether to eat high-fiber cereals, whether to avoid smoked meat.

The press should provide the information and the understanding that is necessary if people are to think critically about decisions affecting their lives. For most people the reality of science is what they read in the press. They understand science less through direct experience or past education than through the filter of journalistic language and imagery. With the exception of an occasional television or radio notice, newspapers and popular magazines are their only contact with what is going on in rapidly changing scientific and technical fields, and their major source of information about the implications of such developments. Good reporting can be expected to enhance the public's ability to evaluate science policy issues and the individual's ability to make rational personal choices; poor reporting is cause for alarm. What, then, is conveyed about science and technology in the press? In 1966 Frank Carey, a writer for the Associated Press, was asked this question. He listed the following news items that had been reported by science writers over the previous 20 years: "the explosion of a nuclear device in Red China . . . the launching of a flying doghouse by the Russians . . . the birth of quintuplets in South America . . . the sex-lure chemical by which the female German cockroach calls her boyfriend . . . the heartbeat of Olga, the whale . . . and such hot potatoes as fluoridation, Rachel Carson versus the bad guys . . . the radioactive fallout from nuclear bomb tests."[3] Today's science journalists continue to report on a remarkable range of subjects. Indeed, they sometimes call themselves the SMEERSHs: "We cover Science, Medicine, Energy, Environment, Research, and all sorts of other SHit."

Science appears in the coverage of dramatic crises, major discoveries, and the feats of science stars. The applications or implications of scientific knowledge, dramatic or unusual events, and technical disputes have the greatest media appeal. A great deal of information about science and technology, for example, is conveyed through the coverage of disputes over technological risks such as hazardous wastes, nuclear power, air pollution, and pesticides. But science also appears in news articles on drugs, food additives, transplants and artificial organs, cancer, and genetic disease. And technical information is even integrated into the news of the day: from descriptions of artificial heart transplants we find out some facts about human physiology; from stories on AIDS we read about research in epidemiology and immunology; from discoveries of chemical dumps, we read about toxicology; from discussions of the Challenger explosion we read about the properties of materials.

But what do we actually learn about science and technology? Consider, for example, the news coverage of developing research on interferon, a protein manufactured in the body when a virus invades a cell. Interferon was discovered in 1953 as a natural therapeutic agent, a so-called "interfering protein" that inhibits infection. The possibility of isolating the protein raised hopes among scientists for the development of a cure for cancer, and this naturally caught the attention of the press. However, the very limited supply of the agent limited progress in the laboratory, and the subject soon faded from public view.

Then in 1975 Mathilde Krim, a politically astute geneticist, organized a major conference on interferon and cancer. Though there were many technical uncertainties about the therapeutic effectiveness of the agent, the conference was intended to publicize the potential of interferon and to win public support for research.[4] Three years later the American Cancer Society (ACS) began to finance costly clinical trials to test the effectiveness of interferon. Both Krim's efforts and the interest of the ACS brought a deluge of press coverage.

The scientific press was careful to qualify the promises of interferon research. *Science*, for example, talked of interferon's "problematic promise," indicating the tentative nature of existing studies, the high cost of isolating the protein, and its therapeutic limits. The popular press, however, was consistently enthusiastic, and interferon quickly developed the public aura of a "magic bullet," a miracle cure for everything from cancer to the common cold.[5]

In 1980 scientists at Biogen, a biotechnology firm, developed a DNA clone for the protein. The ability to synthesize interferon allowed the possibility of producing large quantities at low cost. Uncritically accepting promotional information from a Biogen press conference, journalists welcomed this new technological development as still another miracle. "Like the genie in a fairy tale," the *Detroit Free Press* told its readers, "science came up with the key to the magic potion, a way to produce interferon in bulk."

Reader's Digest talked about a "wonder therapy," *Newsweek* about "cancer weapons" and "the making of a miracle drug." A *Time* journalist built a dramatic story about "the slow, agonizing, painstaking progress of research," the "barely suppressed excitement among medical specialists," and the "staggering implications of research." Other journalists turned their attention to the dramatic increase in stock prices of Biogen and similar firms and focused on interferon as a profitable commodity. *Business Week* described the efforts of researchers at different firms to synthesize the substance economically as a "race" to capture the market: "We have just passed the quarter mile pole and all the horses are in a bunch." *Time* wrote of the "gold mine for patients and for companies." As late as March 1981 the *Saturday Evening Post* claimed that "punters in Wall Street are already laying bets that interferon is a sure winner."

Throughout this period *New York Times* writer Harold Schmeck reported on interferon with caution and care. In early 1979 he wrote that research on the drug was promising but had as yet produced no definitive evidence of its effectiveness. Later that year Schmeck reported on a possible harmful effect of the agent, suggesting that "the seemingly ideal weapon" was less of a panacea than anticipated. In May 1980 he reported on interferon studies that "put cancer use in doubt," once again emphasizing the lack of evidence about interferon's effectiveness and the "modest, controversial, and even negative results of research." He observed that the promise of a scientific advance may raise research money, but also false hopes. In response to this article four scientists from the Sloan Kettering Institute for Cancer Research wrote a letter of protest to the editor of the *New York Times,* expressing their concern that such qualified reporting could undermine public support of interferon research.[6]

By 1982 other journalists began to report information about the toxic side effects of interferon. Scientists had been aware of these effects since the mid-1970s but had provided

no public information for fear it would dampen popular enthusiasm and stall the interferon crusade. But the difficulties
of using interferon as a therapeutic agent became apparent
when four patients treated with interferon in France died in
1982. Then the tone of press reports abruptly changed from
exaggerated optimism to disillusionment: "From wonder
drug to wall-flower"; "If we start promising cures we will
make a tragic mistake." The wonder drug was demoted from
a magic bullet and disease fighter to a mere "research tool."
Newspapers and magazine articles assessed the situation
pessimistically, with headlines such as: "Jury's out on interferon as a cancer cure"; "Studies cast doubt on cancer drug";
"It's a hard row to hoe." Yet most articles continued to suggest a ray of hope: "It may yet become a very important,
perhaps even revolutionary drug"; "There is no limit to future studies."

Several striking features emerge from a review of the
popular accounts of interferon research, features which I
hope to demonstrate in subsequent chapters are characteristic of science reporting in general. First, imagery often re-
places content. Very little appeared in the press about the
actual nature of the research; instead, most articles were
dominated by imagery that appealed to the public preoccupation with cancer and the hope for a panacea. While interferon's short-term usefulness as a therapeutic agent was
problematic, the research did yield significant understanding of basic biological concepts—such as the control of gene
expression in mammalian cells and the regulation of immu-
nity—that will in the long term affect the practice of medicine. But readers following the interferon story would have
learned very little about these important developments.

Second, the press covered interferon research as a series
of dramatic events. Readers were treated to hyperbole, to
promotional coverage designed to raise their expectations
and whet their interest. Inevitably hyperbole led to premature enthusiasm and then to disillusionment. For when

predictions about interferon's curative powers failed to materialize, unqualified optimism in the press quickly shifted to the opposite extreme.

A third feature of science journalism that is revealed by reading the interferon reports is the focus on competition. Scientists and the firms developing interferon were in a "race" for breakthroughs, for solutions. The gradual accumulation of information that is inherent to the research process was not considered news.

Perhaps the most surprising feature to emerge from a consideration of these accounts is the role that scientists played in promoting interferon and in shaping its coverage in the press. Far from being neutral sources of information, scientists themselves actively sought favorable press coverage, seeing public enthusiasm as a means to enhance research support.

These features of science journalism suggest some curious ironies that merit explanation. Although there is more information about science in the press today than there was in the past, public understanding of science and technology is often distorted; this is an age of science journalism, but also science fantasy and scientistic cults. While scientific rationality is valued as the basis of our "knowledge society," science is invested with magic and mystique; we are led to expect "magic bullets" and "miracle cures." While we demand sophisticated science-based medicine, there are widely supported objections to the animal experiments that allow the development of therapeutic techniques; many go so far as to question the value of research. While we welcome technology as the key to progress and the solution to problems, we are increasingly preoccupied with technological risk; we fear the very technologies we most depend upon.

A further irony lies in the ambivalence of scientists toward the press.[7] In recent years scientists have increasingly been seeking media coverage and developing public relations activities toward this end, yet they are mistrustful of journalists and critical of the coverage of their fields. They complain about inaccurate, sensational, and biased reporting that fosters antiscience attitudes. Indeed, as public communication about science has increased, so, too, have scientists' complaints. Journalists themselves are often critical of the way science is conveyed in the press. However, they place blame elsewhere—with their sources of information; that is, with scientists and their institutions. Indeed, while the communities of science and journalism are dependent on each other and both are concerned about improving public communication, relationships between them are strained. As science writer William Burrows describes the uneasy relationship: "Scientists think that whatever they tell a reporter is bound to come out wrong.... Most ordinary reporters would practically cross the street to avoid running into an expert since they consider scientists to be unemotional, uncommunicative, unintelligible creatures who are apt to use differential equations and logarithms against them the way Yankee pitchers use inside fast balls and breaking curves."[8]

•

This book attempts to explain these ironies by exploring what is going on in science journalism. As we read our magazines and newspapers, what do we find out about science and technology, and what messages emerge from the selective news we receive? What characteristics of journalism affect the creation of science news? And how do the public relations efforts of scientists influence the coverage of science in the press?

The popular press is a diverse enterprise. It includes major, nationally distributed newspapers such as the *New York Times,* the *Washington Post,* and the *Wall Street Journal.* Each of these papers has a circulation of over 700,000, but more important, they are influential because they serve as a reference, a bulletin board, for government officials, for reporters who write for other newspapers, and for television.[9] In effect, they establish a tone and a set of standards for journalism. The press, however, also includes myriad local and regional newspapers; most of these papers belong to major chains such as Gannett or Knight-Ridder, each with circulations of over 3 million. Their staff reporting is mainly directed to local issues, and they rely on the wire services for national or specialized news. Thus most of their science stories are picked up from Associated Press (AP) and United Press International (UPI) and edited to reflect local interests. Finally, the press includes the large circulation weekly newsmagazines such as *Time* (about 4.7 million circulation), *Newsweek* (3 million), and *U.S. News and World Report* (2 million), specialized magazines such as *Business Week* (855,000), and science news magazines such as *Discover*.[10]

There are, of course, many differences between local newspapers, with few specialized reporters, and national papers such as the *New York Times*, which employ a stable of experienced science writers who know the science terrain. Yet a surprising feature of science journalism is its homogeneity. While journalistic reports on science and technology vary in accuracy, depth, and detail, most articles on a given subject focus on the same issues, use the same sources of information, and interpret the material in similar terms. For journalists are bound by similar cultural biases and professional constraints. To the extent that they share common assumptions about science and technology, their writing on scientific issues and events takes place within what Todd Gitlin calls a frame; that is, "a persistent pattern of cognition, interpretation and presentation, of selection, emphasis, and exclusion."[11] This frame organizes the world for journalists, helping them to process large amounts of information, to select what is news, and to present it in an efficient form. Their metaphors, descriptive devices, and catch phrases are expressions of this frame.[12]

The journalistic approach to science reporting has varied

over time.[13] The 1960s was a period of scientific and technological "breakthroughs" and "revolutions." Journalists covered the cosmic events of the space program and the dramatic discoveries in the physical sciences with wonder and élan. The frame changed in the late 1960s and the 1970s, as wonder about the marvels of science and technology gave way to concern about environmental and social risks. Journalists shifted their attention from the conquests of science and technology to their consequences, from the celebration of progress to a more critical reflection about the problems brought about by technological change. In the 1980s the technological enthusiasm of the 1960s has been born again though somewhat tempered by the continued fear of risk. The idea of progress has been resurrected as innovation; the celebration of technology has reemerged as a high technology promotion. The old clichés of breakthroughs have reappeared.

These cyclical trends in the press are reflected in the metaphors and other imagery employed in descriptions of science and technology. Although experienced science writers are more self-conscious about language and more restrained than the general reporter ("*We* never use the word 'breakthrough' anymore," science writers tell me), most science reporting shares a style, an imagery, and a particular world view.

Metaphors are a prevalent and important vehicle of public communication in all areas, but they are especially important in communication about science. Explaining and popularizing unfamiliar, complex, and frequently technical material can often be done most effectively through analogy and imagery. But metaphors are more than an aid to explanation; they are also strategic tools. Literary critic Kenneth Burke defines metaphors as "strategies . . . designed to organize and command the army of one's thoughts and images and so to organize them."[14] Similarly, George Lakoff and Mark Johnson insist that a metaphor is not just a rhetorical flourish, but a basic property of language used to define experience and to evoke shared meanings. They suggest that metaphors affect the ways we perceive, think, and act, for they structure our understanding of events, convey emotions and attitudes, and allow us to construct elaborate concepts about public issues and events.[15]

The notion of metaphors, and of language in general, as an active or strategic tool guides this analysis of science journalism. More than simply a source of information about science, the press plays a significant judgmental role. By their choice of words and metaphors journalists convey certain beliefs about the nature of science and technology, investing them with social meaning and shaping public conceptions of limits and possibilities. Was interferon a "magic bullet" or a "research tool"? Was Three Mile Island an "accident" or an "incident"? Was Chernobyl a "disaster" or an "event"? Is dioxin a "doomsday chemical" or a "potential risk"? Is the weakness of science education a "problem" or a "crisis"? Are incidents of scientific fraud "inevitable" or "aberrant"? Some words imply disorder or chaos; others certainty and scientific precision. Selective use of adjectives can trivialize an event or render it important; marginalize some groups, empower others; define an issue as a problem or reduce it to a routine.

Nor are words and metaphors the only way in which journalists convey values. By selecting their stories out of an endless stream of events and issues, they define certain issues and not others as newsworthy. By their choice of headlines and leads they legitimize or criticize public policies. By their selection of details they equip readers to think about science and technology in specific ways.

I approach the study of science journalism from the assumption that public communication is shaped by the cooperation and collaboration of several communities, each operating in terms of its own needs, motivations, and constraints. Journalists and their editors, and scientists themselves all help to influence the presentation of science in the press. The images of science and technology conveyed to the public reflect the characteristics of the journalistic profession, the judgments of editors, and above all, the controls exercised by the scientific community.

Scientists communicate to one another through specialized journals, but they must rely on the media, mainly newspapers and popular magazines, if they want to reach a wider public. Conversely, the press relies on scientists as a source of information about complex but newsworthy aspects of health, energy, environmental, and economic affairs. This mutual reliance plays a particularly critical role in shaping science news. Thus in the chapters that follow I will suggest how science writing reflects the characteristics of both science and journalism as these two professions seek to control the agenda of public communication.

The book will first review, in Chapters 2 through 4, what is reported about science in the press and how it is reported, by exploring the recurring themes and images that are used to describe the work of scientists, the effects of technology, and the problems of risk. It then turns in Chapter 5 to the question of how these messages are received and what impact they are likely to have on the reader and on policy choices.

Chapters 6 and 7 explore the characteristics of journalism that help to perpetuate certain overarching themes and images, and the professional constraints, cultural biases, and editorial pressures that shape the selection of science news. These characteristics of journalism converge with the complexity and uncertainty of many areas of science and technology to reinforce the tendency of journalists to look to scientists as neutral sources of authority. Thus I turn, in Chapters 8 and 9, to the influence of scientists themselves on science journalism as they seek to control the language and content of science and technology in the press. Finally, in Chapter 10, I review some of the fundamental differences that contribute to continuing stress between the two inter-dependent communities of science and journalism.

Scientists employ increasingly sophisticated public relations techniques to assure that their interests are represented with maximum media appeal. Their efforts must be understood in terms of their stakes in a favorable public image— stakes that have developed along with the increase in large-scale and costly research tied closely to applications, the strains in scientists' relations to their traditional sources of funding, and the new patterns of accountability that have resulted from public concern about the social impacts of science and technology.

Control over the information and images, the values and views, the signs and symbols conveyed to the public is an extremely sensitive issue in today's society. Industries, political institutions, professional groups, and aspiring individuals all want to manage the messages that enter the cultural arena through education, entertainment, and above all, the media. Scientists are no exception. Analyzing their relationship with the press is thus a means of shedding light on problems that are of concern to scientists in their quest for improved public understanding of science and greater public support, to journalists seeking to develop a difficult and increasingly important profession, and to those of us who want to be accurately and fairly informed about complex technical matters that affect our lives.

2

THE MYSTIQUE OF SCIENCE
IN THE PRESS

On January 16, 1902, the editor of the *Nation* wrote a critical editorial on the popular appreciation of science. The public conceives of science as a "variant of the black art," of scientists as wizards and magicians, socially isolated from the society. "The scientist appears akin to the medicine man . . . the multitude thinks of him as a being of quasi-supernatural and romantic powers. . . . There is in all this little resemblance to Huxley's definition of science as simply organized and trained common sense."[1] To the extent that the public still holds such attitudes, one should not be surprised given the image of science in the press.

Science often appears in the press today as an arcane and incomprehensible subject, far from organized common sense. And scientists still appear to be remote but superior wizards, above ordinary people, culturally isolated from the society. Such heroic images are perhaps most apparent in press reports about prestigious scientists, especially Nobel laureates. But the mystique of science as a superior culture is also conveyed in the promotion of science literacy, in the coverage of scientific theories, and even in stories about scientific fraud. The result? Far from enhancing public understanding, such press coverage creates a distance between scientists and the public that, paradoxically, obscures the importance of science and its effect on our daily lives.

The Scientist as Star

Each year the press devotes increased attention to winners of the Nobel Prize. Most news magazines have doubled their coverage over the past decade; much of this extra space is being used simply for larger headlines and a greater number of photographs. In the accompanying text, the press, with stunning regularity, focuses on the recipient's national affiliations and stellar qualities, using language recycled from reports of the Olympic games. "Another strong U.S. show"; "Americans again this year receive a healthy share of the Nobel prizes"; "U.S. showed it is doing something right by scoring a near sweep of the 1980 Nobel Prize," "The winning American style." These are the headlines one is likely to find in newspapers across the country.

Just as the papers count Olympic medals, so they keep a running count of the Nobel awards: one year "Americans won eight of eleven"; in another, "Eight Americans were recognized, tying a record set in 1972." "Since 1941," *U.S. News and World Report* announced to its readers in 1980, "the U.S. has had 126 Nobel winners in science, more than double the number won by second place Britain." The stories describe nations as rivals somehow competing in a Nobel race for national pride—an image that obfuscates the international cooperation that is supposed to, and often does, characterize scientific endeavors.

Following the style of sports writing or reports of Academy Awards, journalists emphasize the honor, the glory, and the supreme achievement of the prize: "The most prestigious prizes in the world.... They bring instant fame, flooding winners with speaking invitations, job offers, book contracts, and honorary degrees," runs a typical comment. In 1979 *Time* printed a large picture of the gold medal: "The Nobel Prize Winners are called to Mount Olympus; the recipients have worldwide respect."

Local or regional papers also cover Nobel winners like sports or movie stars but add a twist, seeking to find a local angle, however remote. Consider Roald Hoffmann, who received the 1981 prize in chemistry. A Rochester, New York, paper mentioned that he was a "Kodak consultant." The *New York Daily News* reported that he had graduated from Stuyvesant High School in New York City and printed his picture from the high school yearbook with the caption: "Another example of an alumnus who has done well." A journalist on a Seattle newspaper found a local angle in the fact that one of Hoffmann's Ph.D. advisers was a professor at the University of Washington. Columbia University claimed Hoffmann as one of 39 laureates among its former faculty and alumni. In fact, university publications not infrequently provide counts of the prizes of their alumni and faculty as evidence that their institutions are vibrant and vital research centers. A Harvard publication boasts of "more than twice the number than from any other American university."

In one important respect, though, reports of Nobel awards differ markedly from sports writing. Coverage of sports stars often includes analyses of their training, their techniques, and the details of their accomplishments. However, except in the *New York Times* and specialized science journals, the coverage of Nobel scientists seldom includes details on the nature of the prizewinner's research or its scientific significance. To the extent that research is noted at all, it appears as an arcane, esoteric, mysterious activity that is beyond the comprehension of normal human beings. "How many people could identify with Mr. Hoffman's lecture subject: 'coupling carbenes and carbynes on mono-, di-, and tri-nuclear transmission metal centers' "?[2] The presentation of science as arcane is reinforced by photographs of scientists standing before blackboards that are covered with complicated equations.

In their interviews with journalists, scientists themselves reinforce the mystification of science by emphasizing the extraordinary complexity of their work. Physicist Val Fitch describes his research: "It's really quite arcane." "I find it difficult to convey to my family just what it is I've been doing," said physicist James Cronin in an interview with a *Time* reporter in 1980. *Time* cited the words of a member of the Nobel committee to describe Cronin's work: "Only an Einstein could say what it means."[3]

Just as science is described as divorced from normal activity, so the scientist, at least the male scientist, is portrayed in popular newspapers and magazines as socially removed, apart from, and above most normal human preoccupations. Science thus appears to be the activity of lonely geniuses whose success reflects their combination of inspiration and total dedication to their work. One scientist sees "in a most passive looking object" a "veritable cauldron of activity" that the rest of us are unaware of. Another "stumbled" on his find but then spent a year "probing" for errors.[4] A frequent image is that of scientists spending twelve hours a day, seven days a week, at their work. Reporting on the effect of the prize on Sir Godfrey Hounsfield, one of the 1979 winners in medicine, a journalist from *Time* notes only one change in his life: "He plans to put a laboratory in his living room."[5] Another reporter portrays prizewinning scientists as part of "an inner circle of scientific giants" who talk about science "the way other people talk about ball games." Being with them is "like sitting in a conversation with the angels."[6] That a prestigious scientist can behave like ordinary mortals is noted with an air of surprise. The caption of a *New York Times* photograph of Walter Gilbert (the Nobel Prize winning chemist who gave up his chair at Harvard to run the firm Biogen) notes his managerial skills: "[These] should not be underestimated just because he has a Nobel Prize."[7] Writing of Nobel physicist Hans Bethe's concern about the buildup of nuclear weapons, a *New York Times* reporter remarks that "he ultimately places his faith not in technology but in human beings—a remarkable stance for a man who has dedicated his life to the pursuit of science."[8] The fact that celebrated scientists often teach undergraduate classes or keep office hours is considered a newsworthy point: "Why would a world-class scientist waste time describing electrons to a fidgety mob of 400 students?"[9]

While successful male scientists appear in the press as above the mundane world and totally absorbed in their work, the few women laureates have a very different image. Stories of female Nobel Prize winners appear not only in the science pages, but in such life style and women's magazines as *Vogue* and *McCall's*, which seldom cover science news.

McCall's described Maria Mayer, who shared the physics prize in 1963 for her theoretical work on the structure of the nucleus, as a "tiny, shy, touchingly devoted wife and mother," a woman "who makes people very happy at her home." Approaching the story in a personal style hard to imagine in the coverage of a male Nobelist, the reporter interviewed Mayer's husband, who observed: "She was once a terrible flirt but lovely and brighter than any girl I had ever met." She is, according to this article, "almost too good to be true"; "a brilliant scientist, her children were perfectly darling, and she was so darned pretty that it all seemed unfair." The reporter noted the "graceful union between science and femininity," but also emphasized the conflict between being a mother and a scientist, the guilt, the opportunities missed by not spending more time at home. Writing about Mayer's work, the journalist remarked that she explains it in a "startlingly feminine way" because she used the image of onion layers to describe the structure of the atom. The article then goes on to describe Mayer's reputation as "the faculty's most elegant hostess."[10]

Science journalists used similar stereotypes in their descriptions. A *Science Digest* article called "At Home with Maria Mayer" begins: "The first woman to win a Nobel Prize in science is a scientist and a wife." It showed a picture of her, not at the blackboard, but at her kitchen stove.[11] Similarly, the *New York Times* headlined its article on Dorothy Hodgkin, who shared the prize for chemistry in 1966: "British Grandmother Wins the Prize." And *Time* emphasized her "domesticity" and "elegance of appearance."[12]

The feminist movement did little to dispel such stereotypes. In 1977 Rosalyn Yalow, winner of the prize in medicine, also received extensive coverage in women's magazines. By *Vogue* she was characterized as "a wonderwoman, remarkable, able to do everything, who works 70 hours a week, who keeps a kosher kitchen, who is a happily married, rather conventional wife and mother."[13] *Family Health*, a magazine that reaches 5.3 million readers, headlined an article: "She Cooks, She Cleans, She Wins the Nobel Prize" and introduced Yalow as a "Bronx housewife." The journalist expected to meet "a crisp, efficient, no nonsense type" but discovered that "she looked as though she would be at home selling brownies for the PTA fund raiser."[14]

Journalists had more difficulty fitting Barbara McClintock, recipient of the 1983 Nobel Prize in medicine, into this stereotype, and this in itself became news. *Newsweek* called her "the Greta Garbo of genetics. At 81 she has never married, always preferring to be alone."[15] This article was published in the section called "Transition" (along with the obituaries), although an item on McClintock also appeared on the "Medicine" page. The *New York Times* covered McClintock in a long feature article. Its very first paragraph observed that she is well known for baking with black walnuts.[16]

The overwhelming message in these popular press accounts is that the successful woman scientist must have the ability to do everything—to be feminine, motherly, and to achieve as well. Far from being insulated and apart from ordinary mortals, women scientists are admired for fitting in and for balancing domestic with professional activities. As a remarkable exception to the usual coverage of scientists in the press, the portraits of female Nobelists only highlight the prevailing image of science as an arcane and superior profession, and points up the lack of attention to its substance.

To complete the image of the esoteric scientist, journalists often convey the need of money and, above all, freedom to sustain science stars. A 1979 *Time* article is a typical example, attributing the prominence of American Nobel Prize winners to the "heady air of freedom in U.S. academia and the abundant flow of grants. . . . Just do your own thing, the bounteous government seemed to say."[17] The writer compares this tradition of freedom to the "rigid" British, the "ideological" Soviets, and the "herr doctor" syndrome in Germany and France. Accepting the conventional (but questionable) wisdom among many scientists, he expresses concern that the pressure to apply science to practical ends and to impose "cumbersome" regulations on experimental procedures will limit future triumphs. This reporter also argues that U.S. science gains by insulation from humanistic pursuits: "The best minds have not been overburdened with required studies that are remote from their interests." Scientists, he suggests, do such specialized and important work that they operate outside a common intellectual or cultural tradition.[18]

Ironically, while treating scientists as somehow removed from the common culture, journalists often turn them into authorities in areas well outside their professional competence. Thus we frequently read of their opinions on nuclear power, the arms race, the prospect for world peace. In 1981 *U.S. News* offered its readers the opinions of past and present American Nobel winners on the question "What are the greatest challenges facing the U.S. and the world at large and what can be done to meet those challenges?"[19] The answers, of course, varied, but the message underlying the question was clear: science is a superior form of knowledge, and those who have reached its pinnacle have some special insight into every problem.

Science as a Resource

The image in the press of the scientist as a superstar of knowledge is matched by another image—that of scientific knowledge as perhaps the most important resource of the nation. This view of science as a national resource is most explicitly conveyed in the recent flood of articles and editorials about the "crisis" of science education. In 1983, a *New York Times* editorial defined the problem: "The battle for the future is being waged in the classroom and America is losing. Science and mathematics, once the backbone of education, is now its soft underbelly." This is not the first time that science literacy has been a newsworthy issue. In 1957 the Soviet launching of Sputnik provoked major concern among scientists, politicians, and military leaders about the quality of science education and its bearing on the production of scientific talent and technical skills. The press conveyed this message, and the resulting public concern helped fuel a national effort to improve science education. The resurgence of the cold war military mentality in the 1980s, as well as international competition in high technology, has brought renewed interest in science literacy. As commissions and committees deliberate the problem, their reports have become fodder for the press.

The articles on science education present a generally consistent point of view. Because science and technology are increasingly defining our lives, science literacy is becoming a survival skill. But what is science literacy? For the most part, news articles treat it not as understanding of science, but simply as computer literacy, as "the passport to the electronic universe." Skill in high-technology fields is portrayed as the definition of competence in a competitive society: "A Revolution Is Under Way: The Smokestack Industries Are Shrinking, Leaving Millions without Skills to Compete in the Emerging High-Tech Economy." News articles call resistance to computers "computerphobia," a problem to be surmounted.[20]

Articles on science education emphasize that our school systems are inadequate to the task of training students for the technological age. A recurring word in the coverage of science education is "gap," usually referring to a supposed discrepancy between the skills engendered by the educational system and American industry's need for a technically skilled labor force if it is to compete. A headline in the *Milwaukee Journal* reads "Education Gap Perils Firm."[21] The *San Jose Mercury* claims that this gap "has contributed to America's decline as a world leader in technology."[22] The press throughout the country made much of a statement from a speech on education by Glenn Seaborg, former chairman of the Atomic Energy Commission: "If the deterioration in U.S. scientific competence is not reversed, I think our economic situation is going to continue to deteriorate and we are not going to recover."[23]

Accepting reports of the declining quality of science education, journalists propose numerous explanations, usually related to trends in the 1960s. Thus some blame "the Great Society programs" or the budget constraints that resulted from the shift in national priorities during the late 1960s. Others blame the general "mentality of the sixties," when students demanded a less structured educational system. A *Business Week* reporter writes that "during the politically volatile years of the late 1960s, schools shifted their focus from achievement to social relevance and federal funding for science education was the victim. The public programs of the 1970s drained school coffers," leading to low teacher salaries, out-of-date equipment, and especially a shortage of computers.[24]

Many of the reports on the problems in science education come from corporate sources; the press covers such reports with no hint that vested interests may be involved. *Business Week,* for example, quotes a Texas Instruments executive: "If we don't do something about computer literacy, we will have another kind of haves and have-nots in our society that will be much worse than the black/white division."[25] Just as computer employment ads promise that "career minded employees get ahead and stay ahead," so journalists write that science illiteracy is responsible for leaving many people "behind in the technological dust."

Following corporate warnings, some suggest that inequities in science education will create a "two-culture society," a "gap between the three million professional scientists and engineers and the 224 million others who view technology as an undecipherable and threatening black box."[26] *Time* published an article in 1982 called "The Fuzzies Meet the Techs" about the efforts to bridge the "gulf of mutual incomprehension" between liberal arts students and science students. "The techs are considered by the fuzzies to be nerds. The techs in turn consider the fuzzies as only marginal in reaching logical conclusions." Presenting the techs as superior, the article suggests the importance of teaching fuzzies the techs' way of thinking in order "to overcome technophobia."[27] It says little about the problem of general illiteracy and the importance of liberal education.

By stressing the gap between fuzzies and techs, between computers haves and have-nots, between the needs of society and the availability of people with technical skills, these reports on science education reinforce the mystification of science. And by idealizing technical professions, they oversimplify both the meaning of science literacy and the actual role of science as a national resource.

The Purity of Science

One might think that discussion of incidents of fraud in science, which appear in the press with increasing frequency, would undercut the mystique of the purity of science so prevalent in reporting about prestigious scientists and science literacy. But such is not the case. On the contrary, journalists report deviant behavior in a manner that further idealizes science as a pure, dispassionate profession.

Journalistic interest in scientific fraud developed in the late 1970s as part of the post-Watergate preoccupation with corruption in American institutions. Some of the articles on fraud are in fact similar in style to reports of political or business scandals, describing particular acts of fraud, the investigations that revealed the incidents, and the institutional response. Other articles discuss the issue of fraud more analytically, focusing on the causes and extent of fraud and describing particular cases as symptomatic of deeper problems. These two approaches reflect different interpretations; the first suggests that fraud is simply the deviant behavior of individuals, the second that it is a larger phenomenon with underlying causes that are basic to the present organization of science. Yet both convey a mystique about science, idealizing it as a sacrosanct, if vulnerable, profession.

The first type of news report deals with fraudulent behavior in science as a "scandal," a "betrayal of trust," a "sin against science," a "threat to patients or consumers." Part of the scandal is the reluctance of research institutions to recognize fraud and the inability of responsible authorities to prevent it. For example, a group of investigative reporters from the *Boston Globe*'s "spotlight team" wrote up a four-part series of articles on Marc J. Straus, a research physician specializing in lung cancer at University Hospital, who falsified data on his research subjects in order to show the success of his research on cancer therapy.[28] The headlines of the articles focused attention on both the scandalous aspects of his behavior and the institutional failure to take action despite the seriousness of the offense, which involved human subjects: "Cancer Research Falsified," "Boston Project Collapses," "Doctor under Fire Gets a New Grant." The articles emphasized that corruption was a deviance from professional norms, an unusual event.

The more analytic articles use individual cases of fraud to criticize current research practices. A *New York Times* reporter, in an article entitled "The Doctor's World: How Honest is Medical Research?" calls attention to competitive practices in research.[29] The *Christian Science Monitor* sees fraud as part of "a larger problem," in particular the "corrosive effect of pressure to publish."[30] Other journalists have variously attributed the problem to "the pressure cooker of research," inadequate supervision of younger colleagues, or the fact that most experiments are never replicated because "you don't get a grant for repeating someone else's work." Only a few articles in the *New York Times* raised fundamental questions concerning the validity of certain traditional assumptions about science and scientific method. Can scientific honesty be assumed? Is the scientific method adequate? Does the peer review process offer enough protection against fraud? Most reporters avoid these structural issues, rather describing individual cases of malfeasance as stains on the scientific ideal.

Whether journalists define fraud as an individual aberration or a growing problem in the contemporary practice of science, they project a coherent image of scientific ideals.

The metaphors typically used to describe fraudulent data are instructive. They "contaminate," "tarnish," "besmirch," "taint," or "sully" the reputation of individual scientists, their institutions, and science itself. Faked data must be "expunged," "purged," "withdrawn from the scientific record." Scientists who learn that one of their colleagues is involved are invariably reported to be "shocked," "horrified," "stunned," or "reluctant to believe it." For fraud is a "sin" as well as a scandal. The culprit has "fallen" or "betrayed" the profession. "When a scientist succumbs to temptation and pays the price, it is always sad."

Fraudulent acts in most other fields (except perhaps in sports) evoke quite different and less moralistic metaphors. Consumer fraud is a "ripoff" or a "crime," hardly a sin. Political scandals are abuses of trust and reported, often cynically, as critiques of political institutions. The bribery scandals in New York City appeared as one more example of the corruption inherent in local politics that the press helps to expose. But science, idealized in the language describing scientific fraud, is portrayed as a profession apart—dispassionate, objective, and with values that remain above those in other fields. A *Newsweek* article states this view clearly: "A perception of widespread fakery undermines the trust in others' work that is the foundation of science. More than business or law or politics, science rests on the presumption of honesty in a quest for truth. If that presumption comes into question, a backlash against science may not be far away. And that could compromise what is still one of the more objective and honest sources of information in an ever more complicated world."[31]

The idealization, so evident in the coverage of fraud, has paved the way for the use of science in the press as a neutral source of information for the creation of social policy, and a powerful source of authority in support of popular—if controversial—beliefs.

The Authority of Scientific Theory

Scientific information is often reported in the press, but theories are seldom newsworthy. A notable exception are those theories of behavior that bear on controversial social stereotypes. Thus theories of evolutionary biology and natural selection, when used to explain human differences, have had an active press. The theory of biological determinism attracted considerable news coverage following the controversy over Jensen's claims about the relationship between race and IQ. Its reappearance in the growing field of sociobiology has again attracted the press. The reports on sociobiology have been less concerned with its substance than with its purported applications. In selecting this subject for extensive coverage, journalists are in effect using a controversial theory to legitimize a particular point of view.

Sociobiology is a controversial field devoted to the systematic study of the biological basis of social behavior. Its basic premise is that behavior is shaped primarily by genetic factors, selected over thousands of years for their survival value. Its most vocal proponent, entomologist Edward O. Wilson from Harvard University, contends that genes create predispositions for certain types of behavior and that a full understanding of these genetic constraints is essential to intelligent social policy. He believes that sociobiology is "a new synthesis," offering a unified theory of human behavior. "The genes hold culture on a leash," he writes in his book *On Human Nature*. "The leash is very long but inevitably values will be constrained in accordance with their effects on the human gene pool."[32]

Wilson's arguments about human behavior, extrapolated from his research on insect behavior, have been widely attacked by other scientists for their apparent justification of racism and sexism, for their lack of scientific support, and for their simplistic presentation of the complex interaction of biological and social influences on human behavior.[33] Yet, ever since the publication of Wilson's first book on the subject, *Sociobiology, A New Synthesis,* was reported as news in the *New York Times* and welcomed as a "long awaited definitive book," the press has typically discussed the arguments for sociobiology and the details of particular studies in uncritical, often enthusiastic, terms. Sociobiological concepts subsequently have appeared in articles about the most diverse aspects of human behavior, used, for example, to explain:

- The differences between male and female behavior: "Authorities now say nature not nurture makes him thump and thunder while you rescue lost kittens and crimp." (*Cosmopolitan*)
- Human decency: "Decency is rooted in gene selfishness to enhance the prospect of survival." (*New York Times*)
- Child abuse: "The love of a parent has its roots in the fact that the child will reproduce the parent's genes." (*Family Week*)
- Machismo: "Machismo is biologically based and says in effect: 'I have good genes, let me mate'." (*Time*)
- Intelligence: "On the towel rack that we call our anatomy, nature appears to have hung his-and-hers brains." (*Boston Globe*)
- Promiscuity: "If you get caught fooling around, don't say the devil made you do it. It's your DNA. (*Playboy*)
- Selfishness: "Built into our genes to insure their individual reproduction." (*Psychology Today*)
- Rape: "Genetically programmed into male behavior." (*Science Digest*)
- Obesity: "A genetic tendency to stock for a famine that never comes." (*Science Digest*)
- Aggression: "Men are more genetically aggressive because they are more indispensable." (*Newsweek*)

The press has been most aroused by sociobiology's controversial implications on the subject of sex differences. Not surprisingly, the most uncritical acceptance of the theory appears in *Playboy*. In a somewhat tongue-in-cheek article, called "Darwin and the Double Standard," *Playboy* says the critics of sociobiology are "burying their head in the sand" and "refusing to face facts." The theory, we're told, directly challenges women's demands for equal rights. "Perhaps [women] are defying biology—it's not nice to fool Mother Nature. Recent scientific theory suggests that there are innate differences between the sexes and that what is right for the gander is wrong for the goose."[34]

Such efforts to entertain by playing on conventional stereotypes are not confined to *Playboy*. *Time*, for example, begins an article with the question, "Why do men go to war? Answer: Because the women are watching." The reporter explains that this conclusion is confirmed by sociobiology: "Male displays and bravado, from antlers in deer and feather-ruffling in birds, to chest thumping in apes and humans, evolved as a reproductive strategy to impress females."[35] And *Cosmopolitan*, citing the "weight of scientific opinion" to legitimize its bias, tells its readers, "Recent research has established beyond a doubt that males and females are born with a different set of instructions built into their genetic code."[36]

Cultural stereotypes also attract the press to specific kinds of research. In 1980 two psychologists, Camille Benbow and Julian Stanley, published a research paper in *Science* on the differences between boys and girls in mathematical reasoning. Their study, examining the correlation between Scholastic Aptitude Test scores and classroom work, found that differences in the classroom preparation of boys and girls were not responsible for differences in their later test performance. The *Science* article was careful to qualify the implication of male superiority in mathematics: "It is probably an expression of a combination of both endogenous and exogenous variables. We recognize, however, that our data are consistent with numerous alternative hypotheses."[37] But the press was less qualified, writing up the research as a strong confirmation of biological differences and a definitive challenge to the idea that differences in mathematical test scores are caused by social and cultural factors. The newspeg was not the research, but its implications.

The authors themselves encouraged this perspective in their interviews with reporters, where they were less cautious than in their scientific writing. Indeed, they used the press to push their ideas as a useful basis for public policy. According to the *New York Times,* they "urged educators to accept the possibility that something more than social factors may be responsible. . . . You can't brush the differences under the rug and ignore them."[38] The press was receptive. *Time,* writing of the "gender factor in math," summarized the findings: "Males might be naturally abler than females."[39] *Discover* reported that male superiority is so pronounced that "to some extent, it must be inborn."[40] It was left to a few *New York Times* op ed pieces and to some women's magazines to question the methodology of the research and the limited nature of the evidence.[41]

What is striking about many of the articles on sociobiology is how easily reporters slide from noting a provocative theory to citing it as fact, even when they know that the supporting evidence may be flimsy. A remarkable article called "A Genetic Defense of the Free Market" that appeared in *Business Week* clearly illustrates this slide. While conceding that "there is no hard evidence to support the theory," the author writes: "For better or worse, self-interest is a driving force in the economy because it is engrained in each individual's genes. . . . Government programs that force individuals to be less competitive and less selfish than they are genetically programmed to be are preordained to fail." The application of sociobiology that he calls "bioeconomics" is controversial, he says; nevertheless, it is "a powerful defense of Adam Smith's laissez-faire views."[42]

This journalist and many others writing about sociobiology recognize, indeed rely on, the existence of controversy to enliven the story. Yet most articles convey a point of view by allowing considerable space to sociobiology's advocates and by marginalizing the theory's critics.[43]

In numerous articles, critics of sociobiology are variously dismissed as ideologues, Marxists, feminists, or members of the radical left. They are "few in number but vociferous"; people who are "unwilling to accept the truth." To the extent that their views are presented, they are characterized as distorted or isolated. A *Science News* reporter, for example, wrote that "one runs the risk of misrepresenting the consensus view by focusing, however briefly, on critics and criticism."[44] *Newsweek* suggested that Wilson was a victim like Galileo: "The critics are trying to suppress his views because they contradict contemporary orthodoxies."[45] *Science Digest* compared the criticism of sociobiology to the attack of religious fundamentalism on the theory of evolution—"Like the theory of evolution, sociobiology is often attacked and misinterpreted"[46]—a comparison that places sociobiology's scientific critics, such as Stephen J. Gould and Richard Lewontin of Harvard University, in the same league as William Jennings Bryan.

The uncritical acceptance, indeed promotion, of sociobiology once again reflects the idealization of science as an ultimate authority, albeit selectively applied. For by its selection of what theories to champion, the press in effect uses the imprimatur of science to support a particular world view. It does so, however, with little attention to the substance of science, its slow accumulative process, and its limits.

●

Whether they write about prizes, professional problems, or scientific theories, newspaper and popular magazine reporters convey a sense of awe about science. The scientist is a star engaged in a highly competitive international race for prizes or prestige. Sometimes the intensity of competition in research can lead scientists to fraudulent behavior, but the image of science remains idealized and unscathed. Even when writing about controversial theories that bear on social policy, the press projects an image of science as an esoteric activity, a separate culture, a profession apart from and above other human endeavors. By avoiding the substance of science and ignoring the process of research, the press ultimately contributes to the obfuscation of science and helps to perpetuate the distance between science and the citizen. These effects are highly problematic in an age when science is in fact very much a part of the common culture, integrally tied to public policy and political affairs.

3

THE PRESS ON THE TECHNOLOGICAL FRONTIER

In early 1982 *U.S. News and World Report* looked ahead to developments that could be expected in applied science and technology. The report talked of "breakthroughs" in many areas, of "startling" progress in space age communication, "radical changes" in medicine, "revolutionary" developments in agriculture, and the "far-reaching" effect of emerging technologies on the way all Americans live.[1] The images describing the application of scientific knowledge in this report are characteristic of the coverage of technology in the popular press. In article after article, extravagant claims are made about technological change; each new development promises a transformation of everyday life, whether for good or for ill. Conveyed in these reports is a sense of awe about the power of technology, resembling in some ways the presentation of science in the press. But there is a difference: whereas science appears in the press as an ultimate authority, technology appears as the cutting edge of history, as the new frontier.

The press frequently relocates this frontier, as technological advances occur in different fields or the novelty of an innovation wears thin. In the 1960s space had been the new frontier; the names of the space probes were Pioneer, Voyager, and Explorer. But space flights (until the Challenger accident) soon became routine—often in the news as much because of a misfire or because they include an occasional civilian among their astronauts as for their technological sophistication. In the 1980s frontier images appear instead in the news coverage of computer advances, biotechnology innovations, and the development of new medical techniques.

Every frontier has its dangers, even technological ones. Sometimes, as we shall see, the tenor of press reporting about a new development in technology is apocalyptic, especially when its implications appear to run against the current of prevailing public values or when extravagant claims of its promise are not realized. But the coverage of technology is mainly promotional; the dominant message conveyed is that the new development will give society the magic to cure economic or social ills. What is missing is a clear presentation of the role of technology and a clear assessment of its effects. Perhaps in no areas of technology is this promotional bias in reporting more apparent than in the coverage of computers and biotechnology.

Advertising High Technology

In the early 1970s the phrase "high technology" began to appear as a synonym for computer technology in specialized newspaper articles about investments in the computer industry. Today "high technology" has become a national symbol of progress, much like the space program a decade ago. As one reporter put it, "The term high technology has become as much a part of the political lexicon as motherhood, apple pie and the flag."[2]

Many of the hundreds of articles on high technology celebrate new computer developments as the "dawn of a new era," "the wave of the future," or "the force for revolutionary change." The images imply unlimited progress: "Experts believe that only economics and imagination limit the scope of computer technology: the revolution is real. . . . Every prediction is probably conservative."[3] Computer-based technologies will resolve medical problems and even provide social success; one computer article promotes "high tech ways to meet a man." Articles on, for example, stock prices, product sales, computer camps, career choices, college enrollment, and military strategies all make reference to developments in high technology as the new frontier.

The people who work in high technology are portrayed in these articles as pioneers and armed missionaries. A Minnesota computer consortium is "a pioneer in the midst of the high-technology prairie . . . blazing the trail for the educated around the world."[4] Scientists are "armed" with computers. Sometimes they are "gurus" or "apostles," and their followers, "converts." Boston's Route 128 is "East Mecca," California's Silicon Valley, "West Mecca." Their products are "manmade miracles" or "economic magic."[5]

The language in these articles is one of competition, struggle, and war—an imagery long used to promote science and technology. In November 1957 a cover story of *Newsweek* was called the "World War of Science—How We're Mobilizing to Win It." Science was "the front"; "supremacy" over the "growing army of Soviet scientists" was the goal.[6]

Today the imagery is similar; just as the Soviet Union is labeled an enemy on the political front, so Japan becomes the enemy in the technological arena. "The technological battle with the Japanese," a *Newsweek* article tells its readers, "is really in industrial equivalent to the East-West arms race."[7] The *Newsweek* cover of this issue shows a samurai warrior leaping out of a computer screen.

On the home front, local and regional newspapers typically portray high technology as the answer to revitalizing regional economies, and the inability to compete for high-tech industry as a problem. The *Kansas City Star* characterizes the desperate desire of cities to attract the computer industry: "Like an aging burlesque queen seeking the magic of silicone, the nation's older urban centers are trying for another whirl in the bright lights of prosperity. . . . This Burlesque Queen Needs More Bump and Grind."[8] Utah, according to the *Salt Lake Tribune,* suffers from an "image problem," related to temperance laws, that makes it hard to attract high-technology personnel.[9] However, in Alabama it is not the state's image, but the absence of a major research university, that creates "problems in the mind power race."[10]

Just as the press has tended to follow corporate advocacy by promoting computer education as the basis of science literacy, so it has uncritically adopted the corporate rhetoric in its coverage of high technology. By its frequent promotion of computer applications and its use of corporate sources of information on high-technology products, the press unreflectively accepts the assumptions of an aggressive industry seeking an expanded market. Articles appearing in the *Christian Science Monitor* during 1982, for example, described computers that supposedly provide reliable security systems, build appliances to simulate sight and hearing and even human thought, create art, track the course of acid rain, link prison inmates to a career, offer a way out of the recession, search for people lost in a desert, help officials cope with emergencies, analyze poems and prose, aid the poor, provide hands-on experience in analyzing chemicals, teach complex repair tasks, make complicated concepts interesting to kids, solve crime, aid in the running of a restaurant, turn oil into a renewable resource, and bolster local economies.

The inflated language of press reports on technological developments is strikingly similar to the language of "high-tech" ads in the magazines and newspapers in which these reports appear. Advertisers, for example, sell computers as "the cure for technophobia, the dreaded ego deflator," or as "the latest breakthrough." Frontier metaphors leap out of the computer employment ads: "Join Us in Charting New Territories"; "Technology That Knows No Limits"; "Our Sights Are Trained Just Over the Next Horizon at the Elusive Borderline Where Imagination and Technology Intermingle."[11]

Given the promotion of high technology in the press, it is no surprise that a wide range of other advertisers now employ technology metaphors to convey images of competence, precision, or prestige. For example, automobile ads, often printed on graph paper with engineering cutaways, visually and linguistically convey competence through reference to "laboratory testing," "impressive engineering credentials," and "state of the art research technology." Ads use language drawn from genetic engineering or computer engineering—"biomechanically engineered parts," "computerized, finite elements"—to suggest precision. An automobile music system is not just "high tech" but "higher tech for higher living." Even running shoes are "scientifically designed," reflecting the most "advanced biokinetic research" and incorporating "breakthroughs."

During the 1970s and early 1980s rare was the article in the popular press suggesting that high technology might not be the panacea advertised. By 1983, however, the failures of extravagant claims became defined as news, with reporters referring to high-technology expectations as a "cruel illusion" or "naive exaggeration." In March 1983 *Newsweek* revised its earlier oversell to mock the "visions of Atari Democrats who seem to believe that high-tech companies will be the country's economic salvation."[12] (That February Atari had laid off 1700 workers and shifted its manufacturing operation from Silicon Valley to Taiwan and Hong Kong.)

Some articles began to discuss problems that in fact had been evident all along. For example, a *New York Times* reporter pointed out that high technology can contribute to unemployment, create "mind stunting, mind dulling" jobs, and even encourage new forms of crime.[13] Others complained that high-tech industries may not produce the expected revenues to restore the eroding manufacturing base. In 1983 a Seattle journalist warned his readers: "Despite all the headlines of the past few years, this area will not become Silicon Valley 2 or Route 128 West."[14]

A mere two years later, however, the development of supercomputers—very large-scale computational systems— brought a new round of optimistic and promotional publicity. The press welcomed federally funded supercomputer centers as a symbol of the United States' technological muscle. Few articles questioned the effect that such concentrated allocation of resources would have on other areas of research. Most simply recorded the enthusiastic words of supercomputer advocates, who promise no less than a "second renaissance."[15]

As Star Wars robots compete with samurai warriors in a struggle for technological ascendance, there is relatively little critical analysis of the potential social and economic problems of a high-technology society, the problems of worker displacement, or the limited number of high-technology jobs. The promotional press provides little thoughtful reflection to temper oversell—until, of course, promises fail.

A similar pattern of promotion overlaid by periodic concern characterizes the extensive coverage of biotechnology. Unlike the coverage of high technology, however, the initial reporting on biotechnology was dominated by warnings of potential problems. In the mid-1970s molecular biologists held an international meeting at the Asilomar conference center in California to assess the potential risks of the recombinant DNA research. This was mainly a technical discussion, but the press evoked images of Frankenstein monsters and Andromeda-like strains spreading like an incurable disease. Some reporters worried about "warping the genetic endowment of the human race"; others about "biological holocaust." They characterized the scientists engaged in this research as overly eager and hence slow to acknowledge the possible hazards or the ethical implications of their work. Reporting on anticipated benefits such as genetically engineered "bugs" that would eat oil spills, one journalist, for example, wondered what would happen if the bacteria found their way into petroleum storage tanks. The message? Runaway science needs to be controlled.[16]

Only a few years later, questions of safety ceased to be news. Journalists simply dropped the subject, turning their attention to applications. Techniques of gene splicing, once represented as extremely dangerous, became "a mundane tool," and headlines began to tout the potential applications of the research as miracles.[17] Whereas in 1977 *Time* magazine ran a cover story called "The DNA Furor: Tinkering with Life," three years later it called its cover story "DNA: New Miracle." In 1976 the *New York Times* magazine section published an article called "New Strains of Life or Death?" A 1980 article in the same section was called "Gene Splicing: The Race towards Better Human Health."[18]

In 1976 the national press had given substantial space to the debate in Cambridge, Massachusetts, over plans to develop a recombinant DNA research laboratory at Harvard University, and generally sympathized with the critics of the plan. In 1981 another debate took place, this time over plans to encourage biogenetic firms to locate in the area. The community press covered the controversy, for there was local opposition to the plan. The national press virtually ignored it. A brief note in the *New York Times* business section characterized the controversy as "a debate that threatens to alienate the very industry that the state is counting on for growth in the next decade."[19]

By the early 1980s, then, the "runaway science of genetic engineering" has become another "technological frontier." Occasionally the tone of enthusiasm is tempered by expressions of doubt, sometimes reflecting investor skepticism, sometimes concern about the ethics of manipulating new forms of life. The commercialization of genetic engineering technologies has renewed concerns about the possible risks of releasing genetically engineered microorganisms into the environment, and these are duly reported. But most critical articles appear only in advocacy magazines. A magazine for black readers, for example, covered genetic research as "Tampering with Genes: A New Threat to Blacks," while religious magazines periodically publish articles with titles such as "Life Manipulators" or "Domination over Nature," questioning the limits of science in religious terms.[20]

The prevailing response of the press, however, has been to welcome, indeed, to promote, developments in biotechnology. Reports cover competition for patenting genetically engineered products, the implications of the research for resolving medical, agricultural, and industrial problems with new synthetic products, the proliferation of genetics R&D firms, and, in local papers, the importance of these firms to the regional economy.[21] In the range of newspapers around the country catalogued in *Newsbank*, scientists are described as pioneers, "unlocking the basic laws of nature." "After years of being a dowdy old lady, biology has become belle of the ball." Its revolutionary potential has attracted researchers "in droves," and "bankers [are] in hot pursuit."[22]

Typically, reports on high technology and biotechnology swing from claims of miracles to visions of apocalypse, batting readers back and forth from celebrations of progress to warnings of peril, from optimism to doubt. Often the style and content of reporting may have less to do with what is actually going on than with the promotional efforts of advocates, especially if these efforts coincide with journalists' perceptions of public attitudes and social biases. How journalistic perceptions converge with the promotional efforts of science and technology advocates to shape the news becomes even more evident in the reporting about those technologies that appear to be solutions to social and medical problems. Among the newsworthy technological fixes in recent years have been heart transplantation and various medical interventions such as psychosurgery, estrogen replacement therapy, and in vitro fertilization.

Problems and Promises of the Technological Fix

In most popular press reports of high technology, promotional enthusiasm tends to overwhelm the undercurrent of ambivalence. But the coverage of those techniques that bear directly on problems of health is more ambivalent. Heralded by the press as panaceas and welcomed as affirmation that all problems can be solved, these advances also at times evoke deep skepticism, reflecting cultural, religious, and ethical concerns. This ambivalence is reflected in the coverage of organ transplantation.

Media coverage of transplant techniques began in the 1950s with reports in the press of, in the words of a 1958 *Life* article, a "revolutionary new stage of medicine . . . nearly ready to emerge from the research laboratory." This article, entitled "Science Nears a Goal: Bank of Vital Organs," portrayed transplantation as a technical solution to the most fundamental problems of life and death. The reporter predicted that "when kidneys can be transplanted, two kidneys will be a luxury."[23] A photograph in another *Life* article captured the motto on the door of a tissue bank: "Ex Morte Vita," "From Death, Life." There is "promise of a glowing future," claimed the caption.[24]

By 1968 the glowing future had arrived. Dr. Christiaan Barnard's heart transplant in Capetown, South Africa that year stirred the public's imagination and attracted a flood of favorable and flamboyant reports. Journalists hailed Dr. Barnard's operation as a "surgical landmark," a pioneering venture comparable to space exploration. Barnard became a star; his work was described as a "milestone," a "revolutionary development bound to change our lives." Barely mentioned was the fact that his patient died soon after the operation.

After Barnard's "success," the press heralded America's early heart transplant operations with front-page headlines but few technical details. Emphasizing the heroic nature of transplantation, journalists underplayed the importance of the large technical and surgical team required to undertake complex transplant procedures. Primed by press releases from the participating medical centers, reporters also conveyed the misleading perception that the procedure was a miraculously effective solution for heart patients; they gave little attention to the patients' postoperative histories or their deaths. The transplant was a dramatic event; the aftermath ceased to be news.

By the early 1970s the coverage shifted, and the problems of organ transplantation became as newsworthy as the progress. Headlines announced "Heart Transplant Future Looks Bleak," or "The Tragic Record of Heart Transplants: A New Report on an Era of Medical Failure." The press began to keep a box score of successes and failures of transplant operations, and publicized ongoing debates within the professional community, especially between the two Texas heart surgeons Denton Cooley and Michael DeBakey, who were respectively billed as "Texas Tornado and Dr. Wonderful." Reporters wrote of "transplant furor" and "medical rage," suggesting that the technique was hardly a panacea.[25]

Yet optimistic images once again dominated the coverage of the next technological fix, the artificial heart, in 1982. The first operation implanting the Jarvic-7 model in a human subject, Dr. Barney Clark, took place at the Utah Medical Center, where the hospital staff devoted a great deal of attention to media relations. In effect, this was scientific experimentation in a fishbowl. The center had invited the press, and reporters came, many remaining for the entire 112 days until Clark's death. The medical team included public relations experts who provided the journalists with technical details and even information on medical complications. However, according to *New York Times* reporter Lawrence Altman, much important information that would have served to better educate the public remained undisclosed.[26] Altman wanted to know how the patient was selected and the Jarvic-7 model chosen, and how the Institutional Review Board, set up to handle ethical dilemmas, had entered the deliberations. And he wanted to learn what scientific information was actually gained from the experiment. He argued that public relations control over the release of information limited the reporters' access to such potentially sensitive issues.

For the most part, journalist accounts of the operation were flamboyant and optimistic. They welcomed the human experiment as a "dazzling technical achievement," "an astounding medical advance," "the blazing of a new path," and a "medical milestone."[27] They said that researchers learned a lot from the experiment but said little about what was actually learned. They turned the patient into a hero: "This man is no different than Columbus or the pioneers who settled this valley. He is striking out into new territory."[28] Such frontier images mixed with military metaphors. Thus, technology became a "weapon in the conquest of heart disease." New drugs that facilitated the operation were "weapons in the counterattack" against the resistance of the immune system. The medical technologists were, of course, also heroes: "sleuths of the cardiovascular world," "men who race with death."[29]

Despite their enthusiasm, reporters in the Mormon-dominated region often expressed religious and moral doubts about the operation. They asked, for example: "It has been said that the heart is the symbol of love, site of life, habitat of the soul. Can it be replaced by a simple mechanical pump?" The *Salt Lake Tribune* reporter cited the answer from the director of the hospital's Division of Artificial Organs: "It's true that we may have palpitations, a rapid heart beat when we are in love, but this is secondary. If the owner of the artificial heart would find it pleasant to have these sensations, he can turn up the rate of the pump."[30]

Journalists reported briefly on the hospital's Institutional Review Board, but they had little material available on its deliberations and therefore on how the experiment was evaluated and policed. Thus they simply noted that its members had many sleepless nights while "wrestling" with the ethics of replacing a living heart.[31] Again, by focusing on the heroics of the procedure, the press avoided substantive questions that would inform readers about the process and the nature of the choices involved.

As the operation became more routine, some attention did turn to the high cost of the procedure and the related question of who should receive the heart, but coverage remained enthusiastic. A 1983 *New York Times* article on the conquest of heart disease called the mechanical heart a "breakthrough" and an "astounding medical advance." The attentive reader does learn that aspects of the technology are controversial, that scientific understanding is limited, and that "the road toward the conquest of heart disease still stretches beyond the horizon." Nonetheless, the reporter Harry Schwartz, predicts, "Eventually, we will start clamoring to trade in our forty-year-old hearts at the very first hint of disease."[32]

In 1985 the public was once again deluged with reports, this time of William Schroeder's artificial heart implant at the Humana Institute. This coverage, orchestrated by a professional public relations team hired by the hospital, was more detailed and more technically sophisticated. However, the media mood changed with the short-term ups and downs of Schroeder's health; expressing promises and then dismay, the press provided readers with drama but little perspective.

The view of heart transplantation in the popular press has largely reflected reliance on interested advocates: most press reports have simply regurgitated their claims. The risk of unrealistic promotion of controversial procedures that follows from reliance on such sources was even more evident in the extensive press coverage of quite a different technology, estrogen replacement therapy (ERT). This coverage suggests that promotional efforts to influence the press are most effective when they converge with journalists' perceptions of prevailing social values; in this case, the popular fantasy of remaining forever young.

In 1963 articles began to appear in many newspapers and magazines proclaiming the virtues of estrogen replacement therapy—that is, the use of estrogen drugs to decrease the biological effects of menopause and aging.[33] ERT has been found to allay some of the symptoms of menopause and to reduce osteoporosis, the thinning of bone mass characteristic of postmenopausal women. But the press promoted estrogen theory, promising extraordinary benefits—even miracles. Typical headlines read: "Preventing Menopause," "Science Paints Bright Picture for Older Women." An Associated Press (AP) newswriter cited as fact a scientist's assertion that "there is no reason why they [women] should grow old." In 1964 the *Pittsburgh Post Gazette* promised a new era of youth for aging females. The article termed reluctance to prescribe estrogen "archaic," and claimed that "there is no scientific reason to object to its administration." The same story appeared in many women's magazines, where ERT was touted uncritically as an exciting new discovery, a scientific fact, a cure for growing old.[34]

Clearly the discovery of a pill that would keep women young forever was a newsworthy event. Reports of this discovery not only covered a subject of wide interest, but conveyed a message readers want to hear. The problem with this promotional reporting was that it ignored or underplayed the growing evidence indicating ERT's potential risk.

Who were the experts cited by reporters promoting ERT? The major source of information was Dr. Robert A. Wilson, a gynecologist, an active promoter of estrogen treatment, and the director of the Wilson Research Foundation. Funded by three drug firms, his foundation existed to publish and distribute reports and recommendations about specific products. Wilson had authored a paper in the October 1963 issue of the *Journal of the American Medical Association* that described experiments using estrogen drugs to postpone menopause, as well as a popular book called *Feminine Forever.*

He and other scientists who were advocates of estrogen drugs minimized the risks that were increasingly apparent from cancer studies. They made sure that promotional materials on the rejuvenating effects of estrogen were mailed to newspapers and magazines throughout the country. They succeeded in attracting press attention. Articles promoting the therapy continued to appear despite a growing number of health warnings. An Associated Press report in 1965, called "Pills for Femininity," cited a physician: "It can preserve the femininity of 17 million postmenopausal females in the United States. It would cost far less than a cocktail."[36] A 1966 AP article publicized the claims of an unspecified "medic" who promised eternal youth to takers of the drug; the article criticized doctors who worried about the side effects as "impediments to keeping women young and lovely."[37] *Time* reported that estrogen was a "spring of youth," which would "ward off aging of the skin, breasts, and bones." *Look* in 1966 quoted Wilson as saying that "the vitality and freshness of the young girl need not fade at forty."[38]

In the late 1960s the press cited Dr. Robert A. Kistner, a professor of gynecology from Harvard Medical School, a regular consultant to drug companies, and a writer of popular books on ERT. Kistner was vocally critical of those studies that suggested a relationship between estrogen treatment and endometrial cancer, even in 1969 after the Food and Drug Administration (FDA) had issued warnings to that effect.[39]

The widely publicized U.S. Senate hearings on the safety of oral contraceptives in 1970 called attention to the side effects of estrogen drugs. The press responded, conveying the message that using estrogen for birth control could be harmful for health. But articles on using estrogen to postpone menopause selectively cited evidence to minimize the hazards of ERT and continued to emphasize its promise: "You can stop worrying about the menopause," said *McCall's* in 1971.[40] Still referring to Wilson's 1962 research, *Vogue* in 1974 promised "extra years of vitality," calling cancer worries "needless fear."[41] When a 1975 article in the *New England Journal of Medicine* linked ERT to increased risk of endometrial cancer, this was duly reported in the press, but so too were the persistent claims of ERT proponents. "When we drive the freeways we take a risk," a Beverly Hills gynecologist told Jane Brody of the *New York Times*.[42]

In December 1976 the public relations firm Hill and Knowlton proposed to Ayerst Laboratories, its client and a producer of estrogen drugs, a strategy for offsetting the warnings about the risk of endometrial cancer and for maintaining sales of estrogen replacement drugs. To "restore general perspective," and to "counteract unfavorable publicity" the firm recommended Ayerst contact science editors. When doing so, the firm stressed, "it is important to steer clear of attempting to promote the use of estrogens, and instead concentrate on the menopause. . . . The estrogen message can be effectively conveyed by discrete references to products that your doctors may prescribe."[43]

Coverage of estrogen replacement therapy is only one instance of many in which science reporters have been vulnerable to both prevailing stereotypes and the promotional efforts of advocates of a new technology. Consider, for example, how the reporters covered psychosurgical techniques during the 1940s as compared to how they reported them during the 1970s.

In the early 1940s the press welcomed the new techniques of psychosurgery, then called lobotomy, with uncritical enthusiasm, though prevailing opinion within the medical community remained skeptical of the practice.[44] The press was responding to the promotional efforts of Walter Freeman, a neuropathologist who was actively advocating this dubious technique as a miracle cure for a wide range of mental illnesses. Freeman had performed hundreds of lobotomies. They sometimes produced dramatic changes in emotional responsiveness, and sometimes considerable relief, but the costs in the functional impairment of the patients' intellectual abilities and judgment were often devastating. Given these costs and the enormous variation in the benefits of lobotomy, it hardly could be justified as a routine therapeutic procedure. Yet Freeman solicited science writers to promote the technique. He staged psychosurgery exhibits for them at every American Medical Association meeting between 1936 and 1946. He invited journalists to his hospital to observe the surgery and provided them with stories about the success of the technique and testimonials from selected patients. Journalists in effect paraphrased his words: Lobotomy was "no worse than removing a tooth"; "Surgeon's knife restores sanity to nerve victims." Publicity surrounded psychosurgical operations, with journalists welcoming lobotomy as "a promising cure for intractable problems." The press even reviewed Freeman's medical textbook of psychosurgery as if it were a popular book: "It's not too technical even for a layman. . . . No novelist ever had a more timely subject."[45]

Titles of magazine articles listed in the *Reader's Guide* between 1945 and 1952 suggested that lobotomy was a means of cutting out cares, relieving unbearable pain, check-

ing feeblemindedness, curbing psychosis, helping schizo-
phrenics, solving crime, and curing epilepsy. During this
period the *Reader's Guide* listed 40 articles on lobotomy: 18 of
the titles indicated that its results were promising, 14 were
neutral and descriptive, 4 posed questions ("Kill or Cure?"),
and only 4 titles indicated that lobotomy could cause functional brain damage.

Most *New York Times* articles during the same period
conveyed similar optimism: "A Convict Made Normal"; "A
Possible Preventive for Moral Degeneracy." Of the 18 articles in the *Times*, 11 emphasized the promises of the technique, 3 were descriptive, and only 4 cited those doctors
who were warning their colleagues about the damaging side
effects that must be weighed in extending the use of this
surgery. The potential ethical abuses inherent in psychosurgery were ignored.

With advances in drug therapy in the 1950s, lobotomies
became less prevalent. The practice reemerged in the 1970s,
when the refinement of surgical techniques led to sufficient
professional approval to justify its use on a small number of
very sick individuals. However, at this time the press, reflecting public concerns about abuse, and responding to extreme proposals that psychosurgery could be a "cure" for
overcrowded prisons or even ghetto rebellions, paid attention to the social, ethical, and political implications of
the procedure. The predominant metaphor in the press was
"Clockwork Orange," referring to Stanley Kubrick's popular
film, in which scientists use behavior modification techniques to transform the character of a vicious criminal who,
in effect, becomes an abused hero. The newspeg for many
articles was the story of a prisoner in Detroit who in 1973
agreed to psychosurgery and then backed away. This case
became a vehicle to discuss the ethical implications of the
technique.

The titles of articles on psychosurgery published in the
1970s demonstrate the change in tone. Of the 21 *New York*

Times articles published in 1973 and 1974, 12 were negative, using such words as "mutilation" and stressing the problems of human rights. Seven were descriptive: some focused on the need for regulation, while others described the ethical questions of voluntary consent illustrated by the case of the Detroit prisoner. Two articles weighed the pros and cons of the technique, suggesting the need for better evidence of its effectiveness. None repeated the promises of the earlier period. Of the 21 articles on psychosurgery in the *Reader's Guide,* 12 had negative headlines such as "A Robot" and "Pacification of the Brain."

Like the reporting on estrogen replacement therapy, the press coverage of psychosurgery reflected the convergence of the promotional efforts of advocates and the reporters' vision of prevailing social views. But what happens when views on a subject are clearly divided? In such cases, the press plays up the controversy itself as a newsworthy issue. For example, in vitro fertilization (IVF) has been represented in the press as both a humane technological solution to the problem of infertility and a dehumanizing intrusion of technology. Optimistic announcements of a medical breakthrough were matched by negative images of eugenics. It was "a crack in the door to a Huxleyan vision of a Brave New World."

The birth of Louise Brown, the first test-tube baby, in England in 1978 spawned a deluge of articles describing the event as a "sensational obstetric event," a "major medical achievement," a "miracle of modern medicine."[46] But virtually every article conjured up images of Huxley's baby hatchery. Louise Brown's cry was "heard around the Brave New World," according to *Newsweek.* An article in *Time* referred to "Orwell's baby farm" and to "mass-produced kids." The *New York Times* wrote of the "brave new baby." Often in the same article, journalists covered the event as both a miracle and a sign of the apocalypse. They cited both optimistic physicians ("a very encouraging, happy new tool") and skeptics ("the potential for misadventure is unlimited"), enthusiastic sources ("a dramatic breakthrough, on a par with the discovery of anesthetics and penicillin") and cynics ("It's a cookbook thing, something frogs do in a dirty stream."). They dwelled on the conflicting views of scientists, parents, and even theologians.[47]

The birth of America's first test-tube baby in December 1982 brought another burst of publicity, but with a different slant, clearly influenced by national pride. Welcoming the event as an "American achievement," journalists called it "a test-tube triumph" and "a miracle of love and science" and referred to the infant as "America's baby." Still playing on continued controversy, they now attributed it to pressure from extremists, members of the Moral Majority and right-to-life groups.

Focusing on dramatic achievements or conflicting views, few articles on IVF engage in critical analysis of the real legal, social, and ethical dilemmas involved in reproductive technology: the potential commercialization of childbirth, the problems of long-term embryo freezing are seldom in the news. Rare is the article that deals with the limited success rate of IVF techniques or the high cost of procedures that often fail. Thus what most reports leave out is a realistic assessment of the limits and implications of new reproductive techniques.

•

The press coverage of new technological developments plays on and probably encourages the public's desire for easy solutions to economic, social, and medical problems. Just as high technology is presented as the solution to international competition, so medical technologies are portrayed as solutions to problems of health. Just as exaggerated claims about in vitro fertilization play on hopes of the infertile, so coverage of psychosurgery offers false hope to the mentally ill. Even aging has a technological fix. Similar messages have recently been conveyed in the reporting on AIDS. The press has focused extensively on the search for a vaccine well before this technological solution is in sight, helping to divert public attention from the more immediate need to prevent the transmission of the disease.

As we have seen, this style of reporting often reflects the activities of aggressive sources of information, as well as press perceptions of what readers want to hear. Academic, industrial, and research institutions are eager to promote the latest technologies and therapeutic techniques, and many reporters simply convey their stories of success—especially if they fit with prevailing hopes or beliefs. Thus, failures lead easily to disillusionment, and the result is a tendency toward polarized reporting about technological developments. This tendency has become increasingly evident in the press coverage of technological risks.

4

THE PERILS OF PROGRESS

Dioxin in our groundwater; PCBs in our rivers; fluorocarbons in our atmosphere. These chemical terms once meant nothing to most people. Now they are part of our vocabulary, and we read about them daily in the press.

The promotion of technology in the press has often been tempered by warnings of its perils, as yesterday's technological frontiers come back to haunt the present. In the early 1960s the rapid advance of science and technology brought speculations in the news about the "too-rapid pace of discovery" and the "mixed blessings of science." Ten years later the environmental and consumer movements focused these speculations on the potential risks to human health and survival posed by the products of technology.

The reporting of risk is a difficult area of science writing.[1] Journalists must cope with complex and uncertain technical information and sort out conflicting scientific interpretations. Definitions of risk, as we have seen in the coverage of psychosurgery, often rest on prevailing social or cultural biases. Norms of objectivity and fairness encourage reporters to balance different views—to give a technology's critics and proponents equal time—but such efforts expose them to criticisms from all sides. Industrial interest groups and some scientists accuse reporters of taking a biased, sensational, antitechnology approach to reporting risks; they blame the press for creating unwarranted fear of technology and mistrust of industrial practices. Other scientists, environmentalists, and consumer advocates accuse the press of relying unfairly and almost exclusively on "establishment" expertise and of burying stories that might challenge local industries.

To understand some basic characteristics of risk reporting we will look at the extended press coverage of the controversies over the effect of fluorocarbons on the ozone in the atmosphere, the ban on artificial sweeteners, and the health effects of dioxin in waste disposal dumps. The coverage of these issues emphasizes their controversial aspects— the competing interests, disputed data, and conflicting judgments about the hazards of technology. Often the press portrays science as the arbiter or judge of technological perils. Scientists, even when interested parties in disputes, are viewed as the source of authoritative evidence and definitive solutions. Yet, with some notable exceptions, we seldom read about the scientific issues involved in risk disputes or the methods of risk analysis. Thus we are left with no basis for making meaningful judgments about competing claims.

The Ozone Controversy

The ozone controversy in the 1970s attracted considerable media coverage, for at stake was potential atmospheric disaster as a result of the chlorofluorocarbons used as propellants in aerosol spray cans and coolants in refrigeration. The release of these chlorofluorocarbons into the atmosphere, scientists claimed, was depleting the layer of ozone which shields the earth from harmful solar radiation. Press reports on "the spray can war" that followed this claim exposed the public to a scientific dispute over an issue few people had ever heard about before.

In 1973 British chemists published a paper in *Nature* expressing concern about the release of fluorocarbons into the atmosphere and their possible damage to the ozone layer.[2] A Swedish science journalist wrote a news article on the paper after hearing it delivered at the Royal Swedish Academy of Sciences, but newspapers in the United States failed to pick up on the story. The *New York Times* and *Washington Post* avoided it deliberately, believing that it was one more doomsday report with little evidence to support it.[3] Then in 1974 two University of California chemists, Mario J. Molina and F. Sherwood Rowland, described their similar findings in *Nature*[4] and subsequently presented their data at the September meeting of the American Chemical Society. Just before the meeting, Dorothy Smith, the news manager of the American Chemical Society, flagged the paper as a top news story and alerted the media by organizing a press conference on the fluorocarbon problem. The story took off when Walter Sullivan wrote an article for the front page of the *New York Times* on September 26, 1974, explaining the dimensions of the problem. Publishing first, the *Times* in effect defined the assumed risk as news, and over the next year articles followed in newspapers throughout the country.

Much of this press coverage was apocalyptic and sensational, suggesting global catastrophe. The *Philadelphia In-*

quirer wrote that "the world will end, not with a whimper but with a quiet p-s-s-t. . . . The earth may have already committed partial suicide or at least severe self-mutilation." The *New Haven Register* described an industry spokesman in the following terms: "Beneath the guise of Kris Kringle affability lurks one of the men who may cause the end of our present day earth." A widely published AP dispatch asked, "Is a homely aerosol spray can and its charge of propellant gas sowing the seeds of doomsday?"[5]

Chemical industry executives responded, criticizing the press as a "dupe of environmentalists, consumerists and anti-industry scientists" and debunking the evidence from the Molina–Rowland study. The *Wall Street Journal* quoted an industry adviser as saying that the ozone depletion theory "resembles a structure held together by Tinker Toys, Scotch tape, and rubber bands."[6] In the face of an increasingly negative press, Du Pont and other producers of propellants organized a public relations program aimed specifically at journalists. Reporters were inundated with press kits, "clarifications," and public service advertising with leads such as "The Ozone Layer Versus the Aerosol Industry—Du Pont Wants to See Them Both Survive." Du Pont hired a British scientist to tour the country rebutting critical studies.

Reporters reacted to this public relations effort with suspicion. They called the touring British scientist a "scientific hired gun." *Business Week*, however, quoted a chemist with approval: "If a ban were declared on fluorocarbons, we would have to go back to delivering blocks of ice to refrigerate food."[7] The warning sounded much like the message from the nuclear industry, that if nuclear power is not developed, "we will all freeze in the dark."

In 1976 some new data underscored uncertainties about the extent of ozone depletion. The ensuing controversy among scientists led to confusion in the press, as journalists struggled to understand and report on the rapidly changing and controversial nature of the technical arguments. A long-awaited report by the National Academy of Sciences (NAS) contributed to the confusion. The Academy report confirmed the scientific studies warning that chlorofluorocarbons cause harmful deterioration of the ozone layer, though it admitted there remained uncertainties about the severity of the effect. But, concerned about the economic repercussions on industry, the report's authors cautioned against the imposition of regulation until there was further research.[8] The ambiguities in this report left room for widely differing interpretations by the press. Thus, for example, the *New York Times* headlined its account of the report "Scientists Back New Aerosol Curbs to Protect Ozone in Atmosphere," while the *Washington Post* headlined its article "Aerosol Ban Opposed by Science Unit."[9]

Some inconsistencies in the coverage followed from efforts to give stories a local slant: In Orange County where Rowland, one of the major fluorocarbon critics lived, for example, a local paper ran a story on the NAS report under the headline "Aerosol Spewing Earth's Death," while a paper in Wilmington, Delaware, home of Du Pont, reported that "Ozone Study Gives Du Pont a Reprieve." But most of the inconsistencies throughout the coverage of the ozone dispute reflected the waffling of scientists in response to rapidly changing evidence and their preoccupation with the economic impact of their recommendations.

Confused by complexity, most reporters on the subject chose to remain silent on the nature of the evidence and the substance of the scientific dispute. Instead they paid homage to the technical uncertainty simply by balancing the views of contradictory sources. By balancing disagreements without dealing with their substance, they gave the reader no way of judging the validity of conflicting claims. However, the very existence of a publicized dispute had its effect. By late 1976, according to a public attitude survey conducted for the Consumer Product Safety Commission, 73.5 percent of the public had heard of the ozone issue, mainly via the press, and more than half had decreased their use of aerosol cans.

Then, in 1978, the Environmental Protection Agency imposed a ban on the use of chlorofluorocarbons as aerosol propellants. For the press, this ended the problem; no longer controversial, it was no longer news. However, the problem persists, for the compound is still used extensively as a coolant in refrigeration and automobile air conditioners, as a foaming agent in the manufacture of polyurethane, and as a component in industrial solvents. Without press coverage, the public is uninformed and therefore apathetic about a problem which is increasingly severe. Thus industry has been able to block further regulation with no opposition.

The preoccupation with the existence, and not the substance, of controversy has also characterized the press coverage of the carcinogenic risks of food additives, in particular artificial sweeteners. The reports on the sweetener dispute illustrate the difficulties of dealing substantively in the press with the scientific basis of controversy and the resulting reliance on sources interested in biasing the news.

The Sweetener Dispute

"Bitterness about Sweets," "A Fatter Outlook for Diet Food," "Sweeteners Take Their Lumps"—press coverage of controversies over artificial sweeteners is spiced with puns (often in poor taste), but it is also filled with images (often misleading) of science and its role in risk disputes. Cyclamates in 1969 and saccharin in 1977 were found to cause cancer in laboratory animals. The 1958 Delaney Amendment to the Food, Drug, and Cosmetic Act, required the Food and Drug Administration (FDA) to ban any food additives that were found in laboratory tests to induce cancer in animals or man. But the potential cost to the $2 billion a year diet food industry and the $500 million market in food additives turned proposals to ban sweeteners into major public disputes.

Two issues dominated these disputes: how to establish conclusive evidence about risk and whether to ban the food additives immediately or wait for definitive evidence of harm to human health. The press dealt extensively with the role of science as a basis of regulatory decisions and with the validity and significance of the scientific tests used to evaluate risk.

The images of science conveyed in the press coverage of the sweetener controversy were similar to those conveyed in the ozone dispute. Metaphorically, sweeteners were compared to suspects in a crime, under suspicion of causing cancer, and scientists were either detectives investigating the allegations or judges trying to reach a verdict. Such images clearly implied that science would resolve the controversy. Scientists would discover the truth about the carcinogenicity of artificial sweeteners, and the facts uncovered would guide policy decisions and be acceptable to all interests. In October 1969 *Business Week,* for example, announced that the "battle royal over the safety of artificial sweeteners is expected to be resolved next month," and that the food industry was waiting for the "definitive study."[10] Ten years later, the reader was still waiting for definitive answers, and *Newsweek,* now referring to saccharin, announced an eighteen-month "crash effort" by scientists to decide whether it did indeed cause cancer. The results of this effort, reporters implied, would "resolve the dilemma of food additives like saccharin once and for all."[11]

This characterization of science perpetuates several popular myths: that science can provide definitive answers about risk, that "facts" speak for themselves rather than being open to interpretation, and that decisions about what risks are socially acceptable are scientific rather than political judgments. However, this image of science as the ultimate arbiter of the sweetener controversy was implicitly placed in doubt by the frequent coverage of public disputes among scientists over the risks of sweeteners and the need for a ban. During the 1972 congressional hearings on cyclamates, for example, the press reported on debates between FDA Commissioner Charles Edwards and Samuel Epstein, professor of environmental medicine at the University of Illinois, over the need to enforce the Delaney Amendment. In 1977 journalists reported the disputes between Sidney Wolfe, from the Health Research Group, and Kurt Isselbacher, a Harvard professor of medicine, over the costs and benefits of the cyclamate ban.[12]

Drawn to these debates by their interest in controversy, journalists raised questions about scientific objectivity. They identified the participating scientists with their vested interests in the regulatory decisions at stake. *Business Week*, for example, criticized the "environmental mutageneticists" for using cyclamates as a "ready platform" for their broader interests. *Time* referred to "Hertz-rent-a-scientists" who were supported by industry.[13] Thus, on the one hand, science appears in the press as a necessary guide to policy, while on the other, it is described as a politically biased institution strongly influenced by vested interests. While both contain elements of truth, juxtaposing attacks on scientific objectivity with images of science as a neutral arbiter conveys confused and contradictory messages.

The coverage of a policy report on saccharin and food safety prepared by the NAS's Institute of Medicine is an example of the confusion that frequently characterizes reporting on risk. The report, issued on March 2, 1979, recommended a flexible system of regulating food additives. The 1958 Delaney Amendment, automatically outlawing food additives that were found in laboratory tests to induce cancer, seemed potentially too rigid in view of dramatic improvements in cancer detection techniques. The system proposed by the NAS was intended to provide the basis for a new law that would permit the FDA to assess the nature of the risks and to weigh them against benefits when deciding whether or not to ban a substance. Using the word "risk" to describe both the frequency of harm (i.e., the chance of harm from the substance) and the severity (the extent of possible harm), the NAS report placed saccharin in a "moderate to high risk" category of food hazards. It also emphasized the lack of evidence supporting the dietary benefit of saccharin that could be weighed against the risk. Nevertheless, the NAS panel stopped short of recommending a saccharin ban.

This report, like the NAS report on ozone several years before, was thus ambiguous both in its special use of the word "risk" and its waffling on policy recommendations. Not surprisingly, its ambiguity, in the words of science journalist Daniel Greenberg, became "red meat for the carnivores of the news media."[14]

One day before the report was released, the *New York Times*, obtaining its information "from sources familiar with the forthcoming report," described the debate among panel members as a "standoff."[15] The article went on to note the judgmental and emotional factors that enter into scientific evaluation when it serves the needs of public policy, and emphasized—some panel members later said, overemphasized—the extent of disagreement within the group.

The *Times* also interpreted the panel's language (placing saccharin in a "moderate to high risk category") in lay terms to mean it was a "moderate cancer-causing agent." The scientists on the panel strongly objected; they understood their technical terminology to refer to both the frequency and severity of potential harm, not the likelihood that saccharin would cause cancer. The confusion over language was understandable, but it became a point of tension, exacerbated when the *Times* interpreted the academy recommendation as favorable to industry: "People who feel life would hardly be worthwhile without calorie-free sweeteners can probably relax."[16]

Readers of the *Washington Post* also had little opportunity to gain a realistic picture of the findings and their implications. A March 3 news article cynically observed that, although the panel found saccharin to be a "high or moderate risk compound," it did not recommend a ban.[17] On March 5 the *Post* published a column by the president of the Calorie Control Council (the industrial trade association of the sweetener industry), claiming that the academy found "no evidence" of any association between saccharin use and cancer. A few days later the *Post* published still another view, featuring a statement by Fred Robbins, the chairman of the panel, who suggested that saccharin is a potential cancer-causing substance and should be phased out in three years.

This extensive public communication about the saccharin report did little to enhance public understanding of risk. Neither scientists nor reporters effectively communicated the issues underlying the debate; that is, the problems of regulation under an outdated law or the difficulties of balancing costs and benefits for purposes of regulation. For the most part the press simply accepted the NAS report uncritically as support for avoiding a permanent saccharin ban, and did not adequately deal with the challenge to the Delaney Amendment.

During the debate on artificial sweeteners, the diet food industry, represented by the Calorie Control Council, had engaged in aggressive advertising and public relations efforts to influence the press. In a widely circulated brochure the council claimed that the NAS urged Congress and the FDA not to ban saccharin, and encouraged consumers to exert political pressure against "penalizing millions of diabetics and weight-conscious Americans." Citing phrases from the report out of context, the council's brochure also claimed that the academy "highlighted the benefits of saccharin," when in fact it had only listed the claimed benefits in order to show the limited evidence supporting them.

In its public relations efforts the council also took advantage of the public's technical innocence about the validity and usefulness of the animal tests that were used to evaluate carcinogenicity. The carcinogenicity of saccharin had been discovered in 1977 when a researcher, using the standard experimental procedure, fed his rats a dose equivalent to 50 times the maximum level considered safe for human consumption. The rats developed bladder cancer. Toxicologists routinely use such animal tests to clarify the effect of various toxic substances. Because it would be prohibitively expensive to use hundreds of animals in a long-term, low-dose test, it is standard practice to administer high doses in order to increase the likelihood of obtaining statistically valid results; lower dose effects are then extrapolated from the data.

The council debunked this scientific procedure, and many reporters followed suit without discussing the nature of animal testing. Relying on the council's public relations materials, they described the research as "ridiculous," "incredible," and "absurd." Some popular magazines compared the dosage to that of an adult consuming 300 ten-ounce bottles—24 gallons—of dietetic soft drinks per day. Others scoffed at a dosage they claimed to be equivalent to drinking 1250 cans of diet soda a day. *Newsweek* quoted without comment a spokesman from the William Wrigley company: "They give enough to sink a battleship and call it dangerous."[19] *Newsweek* did mention that administering large doses is "standard procedure." But even here the rationale for the procedure was not addressed, and the article went on to quote a politician who lost 47 pounds on a diet supplemented by low-calorie sodas: "People would die of gas before they would die of cancer" if they drank that much.

While expecting answers from science, the press did a poor job of explaining scientific procedures and concepts, including the validity of animal tests and the nature of scientific uncertainty in risk analysis. The idea that large-dose testing might be a legitimate method for determining carcinogenesis was virtually absent from the press coverage; yet, inasmuch as human subjects cannot be used, such testing remains an important research tool.

Assuming the industry position, reporters then used the

FDA's acceptance of large-dose testing to question its credibility. Under the Delaney Amendment, the FDA had been forced to place a temporary ban on saccharin, but the press suggested that in doing so the agency was "arbitrary, ill-organized, arrogant, inbred and inept."[20] As a focus for press criticism, the FDA played much the same role in the sweetener dispute that the Environmental Protection Agency (EPA) would play during the debate over dioxin only a few years later. In both cases, the issue of agency competence reflected the growing concern in industry and the press about appropriate regulation of science and technology.

The Dioxin Debate

A syndicated cartoon pictures Noah's ark stranded on a knoll. An animal on the ark asks his mate: "Why can't we leave now that the flood is over?" The answer: "Dioxin." Press interest in risks associated with the disposal of toxic chemicals, especially dioxin, began with the controversy over Love Canal, Hooker Chemical Company's abandoned waste disposal site in upstate New York. Hooker had dumped about 21,000 tons of chemical wastes in this canal between 1942 and 1953, then covered the dump and sold it for a token payment to the Niagara school board. A neighborhood developed on the site. The local press, in a town heavily dominated by the chemical industry, for years downplayed the occasional complaints from residents about chemicals leaking in their yards. But in 1976 some enterprising investigative reporters on the *Niagara Gazette* opened the issue to public scrutiny by publishing reports of heavy concentrations of chemicals in the area.[21]

The attention of the local press prompted a state investigation and a health alert, which closed the school, evacuated some residents, and encouraged the formation of a citizens group of local homeowners. Love Canal then reached the national press, reported as both a local disaster and a symbol for the growing problem of toxic waste disposal. The human interest angle of a child's death and a family's anguish, the David versus Goliath drama of local homeowners battling against state and federal bureaucracies, the anomaly of scientists scrapping in public about the severity of risks, the dramatic images of deformed babies and forbidden zones, and then the realization of the national dimensions of the issue all combined to make Love Canal a newsworthy event. By October 1981 the press reported that 114 sites in the United States had been classified as dangerous toxic dumps. A year later the number of such sites grew to 418.

This was the situation when dioxin was found in Times Beach, Missouri, in 1982. Complaints about this Missouri dump had first been brought to the attention of journalists nearly a decade earlier, but with little awareness of the problems of toxic waste disposal the complaints were not defined as news and gained neither press publicity nor official attention. In 1980 a Missouri reporter tried to cover the issue but was told by his editor to let the wire services handle it. Even the 1980 discovery of illegally buried toxic waste drums was essentially ignored.

In August 1982 the EPA informed residents of Times Beach that their property was contaminated with dioxin. This was news, but it was reported mainly as a local problem. Then, a few months later, a UPI correspondent published an internal EPA memo indicating there were many potential dioxin sites in the state. Suddenly the issue took on new dimensions; it began to receive increasing national press coverage, clearly placing the risk of dioxin on the public agenda as a problem that could no longer be ignored.

What images were conveyed by the press in the course of the extensive coverage of dioxin? At one extreme, a number of sensationalized reports have portrayed dioxin as another peril of technological progress. A 1980 *Time* cover story was called "The Poisoning of America." The cover showed a man swimming in a pool. Those parts of his face and body that were under the surface had dissolved to leave only the skeleton.[22]

The *Los Angeles Times* in 1983 headlined an article "Lethal Dioxin Monster Guest in Chemical Lab." "Modern wizards of science with ultrasophisticated research tools," the reporter wrote, "have invented headache remedies, dandelion killers, and assorted bug poisons. They have also created a chemical monster." The reporter went on to compare Dow Chemical scientists to horror movie chemists who accidentally create hideous mutants. Dioxin, he claimed, is "classified by scientists as the most deadly compound in nature after botulism and tetanus toxin."[23]

At the other extreme, a 1983 piece in the *Milwaukee Journal* described the dangers of dioxin as "overstated" and "unwarranted." "Is dioxin deadly?" asks the journalist. "Yes . . . for guinea pigs . . . and as far as is known now, not for humans." The article referred to the "partisan politics and unsophisticated reporting" that have caused "hysterical coverage." It quoted a "weary scientist" who did not want to be identified: "Dioxin is bad, dioxin is evil, dioxin is Darth Vader. Saying something good about dioxin is like saying Richard Nixon was a good father. It may be true but nobody would believe it."[24]

Reports on the health effects of dioxin have tended to dwell on the technical uncertainties that confound understanding of long-range risks. As in the coverage of artificial sweeteners, risk appears as a mystery to be resolved by a court of scientists. Dioxin is the suspect, the culprit; science is the judge. But many of the scientists cited were from the industries involved. In an *Atlanta Journal* piece called "Dioxin Studies Under Way but Verdict Is Not In," the journalist explained the differences between the effects of dioxin on animals and the potential effects on humans, citing Dow Chemical scientists: "The human problems from exposure have not been documented."[25]

Unsolved "mysteries" about the effect of dioxin were also the theme of a 1983 *Arkansas Gazette* article: "No pattern, only new questions were found in research on dioxin, Dow experts say."[26] This article also quoted Dow Chemical experts (on tour as part of the firm's $3 million public relations campaign) on their doubts about the relevance of animal tests for assessing dioxin's effect on human health.

Again reporters have tried to deal with uncertainty about risk by balancing different points of view but have provided little by way of critical analysis that would help readers weigh their cogency. A *Time* magazine article in May 1983, for example, quoted a Dow Chemical scientist ("no evidence"), the epidemiologist Irving Selikoff ("no question about it, dioxin is harmful to humans"), and a local housewife near a dioxin site ("almost everyone has thyroid problems"). Judgments of danger from those who were quoted ranged from "no risks" to "Armageddon."[27] (Interestingly, of the hundreds of articles about the health fears of community residents near dioxin sites, few have dealt with the workers in the chemical companies, who face similar risks every day.)

The most important issue for the press became the credibility of the state and federal agencies responsible for regulating science and technology—especially the EPA. The press had long ignored the deteriorating situation within the agency, but in 1983 a series of scandals over the withholding of information about dioxin contamination focused attention on the governmental failures that contributed to the toxic waste problem. Of the 82 *New York Times* articles on toxic wastes in the spring of 1983, 12 focused on industry practices, 45 on the EPA. The articles on industry reported cases of subterfuge or malfeasance: coverups of known risks, illegal practices, closed meetings, or reluctance to cooperate with regulation. But they raised no structural questions about the nature of industrial practices that contributed to the dumping of toxic waste, or more generally to the risks of technological development. Thus the problem appeared as an aberration, a serious but limited problem of regulatory control. Those *Times* articles that focused on the EPA, however, were far more critical, attacking the politicization of the agency, its collusion with industry, and its competence to deal with a confused and disruptive problem.

●

The newspaper coverage of dioxin, saccharin, and other risk events has been called hysterical, sensational, biased, and confused.[28] Clearly some reporting of risk is sensational, and many articles on this complex subject are confused and misinformed. But what are the sources of these problems?

The confusion, characteristic of some though by no means all risk reporting, in part reflects the journalistic response to fast-breaking events, technical confusion, and conflicting information. Reporters have little time to do the research necessary to make an independent and probing analysis of rapidly unfolding events. Thus they rely on available technical sources of information. But these sources often perpetuate the confusion. As we saw in the cases described above, scientific disagreements reflecting the state of technical knowledge about health or environmental effects contribute to confusion, as do the efforts of scientific panels to balance health risks against the economic costs of regulation. So too do the efforts of interested groups to project their own interpretation through the press. Applying naive standards of objectivity, reporters deal with scientific disagreement by simply balancing opposing views, an approach that does little to enhance public understanding of the role of science.

Though some of the methodological issues of risk analysis require little technical background, few reporters try to explain to readers the nature of evidence necessary to evaluate human risk and the problems of judging how much evidence is necessary to warrant policy intervention. Perhaps more important, the press tends to reject statements by scientists who try to explain that they themselves do not know the extent of risk. In business to root out stories and expose potential coverups, reporters and their editors want definitive answers. They suspect that those scientists who claim lack of knowledge are trying to withhold information, to maintain secrecy, or to give them a "runaround." Seeking order and certainty, they convey the idea that science is a solution to problems of risk, that "the acid test of scientific review" will resolve all uncertainties. But in doing so, they perpetuate a false image of science, its contributions to the resolution of risk disputes, and its limits as a basis for public policy decisions.

5

MEDIA MESSAGES, MEDIA EFFECTS

Journalists often cast the problems of technology in the form of a myth or social drama. Communities (Times Beach or Love Canal) are threatened by evil. Normal institutions (departments of health) fail to deal with the threat. Villains (EPA administrators) are identified, replaced, or brought into line through redressive action. Institutions (chemical companies or the EPA) are responsible for restoring order by cleaning up. Solutions are sought in better technology (chemical waste containment facilities) or through scientific knowledge (expert advisory panels). The message is our ability to win over the forces that besiege us. Order is restored.

Prevailing values of order and efficiency shape the press coverage of many areas of science and technology. Scientific theories explaining behavior are uncritically promoted in the press when they reaffirm comfortable social values. Scientific fraud is described as a violation, a moral affront to the purity of the social system of science. Similarly, the risks from chemical waste disposal facilities, pesticide plants, or nuclear power stations are reported as problems of corporate or governmental negligence in a structurally sound system. Accidents, whether of space vehicles or in industrial plants, are blamed on aberrant individuals, sloppy corporate practices, or unfortunate decisions,—then euphemistically labeled "human error." They seldom foster much discussion of deep-seated structural dilemmas—fundamental problems in the organization and regulation of industry or distortions in the allocation of resources—that can lead to neglect of public safety. Interestingly, after the Chernobyl accident the press did raise systemic questions about centralized control and its implications for safety in the Soviet Union; but such discussions seldom take place in the American context.

Ideological assumptions also lead many science writers to adopt the language of cost-benefit analysis without questioning the value of efficiency as a measure of good social policy. Assumptions about efficiency, for example, guide reporting on high technology, where the press often serves as a mouthpiece for the computer industry, giving little attention to the social costs of technological change.

Except for the brief period in the late 1960s, press coverage of science has been shaped by the widely held belief that science is distinct from politics and beyond the clash of conflicting social values. Although individual scientists are sometimes criticized as biased, science as an institution is assumed to be a neutral source of authority and a basis for just solutions in controversial public affairs. Science writers reinforce this assumption, seldom analyzing the distribution of scientific resources, the social and political interests that control the use of science, or the limits of science as a basis for public choice.

The images of science and technology in the press, as we have seen in Chapters 2 through 4, are generally positive but to some extent polarized. Continual progress—new fixes, new devices, new cures—are promised and yet, not infrequently, today's exaggerated promises become tomorrow's sensationalized fears. What is the effect of such polarized images? How do readers respond? And does science journalism in fact shape, or even create, public attitudes, rather than simply mirroring them? It is, after all, the possibility of significant media influence on attitudes and policies toward science and technology that has stimulated the growing concern about the images that appear in the press.

Media Influence on Public Attitudes

Nineteen seventy-nine was a year of widely publicized technological failures: Three Mile Island, the falling of Skylab, the crash of a DC-10, and a series of automobile company recalls of defective parts. In July of that year a group of industry representatives, academic scientists, engineers, social scientists, and pollsters met at MIT to consider the impact that extensive press coverage of these events was having on the public.[1] Opinions among the participants varied widely. Nuclear industry representatives believed that "an exaggerated and irresponsible press" had delivered the industry "a devastating blow." They felt that journalists were "smearing science" in their reporting of engineering accidents. Public attitude surveys, however, indicated no significant change in behavior or loss of public confidence in science and technology. Rather, they suggested the remarkable continuity in the public's trust; despite short-term fluctuations, public attitudes toward science seemed to remain, in the words of one participant, "overwhelmingly favorable."

Again, in the mid-1980s a series of technological disasters—the Bhopal chemical spill, the explosions of the Challenger space shuttle and the Air Force Titan rocket, and the fire at the Soviet nuclear power reactor at Chernobyl—were extensively reported in the press. The critical coverage could have provoked a sweeping Luddite response, but after the initial shock, public attitudes appeared to favor further development of the technologies, except for nuclear power.[2]

The influence of press coverage on public attitudes is difficult to assess.[3] In the 1920s and 1930s, communications research focused on the media's power to shape public opinion and, in particular, to impose political ideas. Thus, Charles Cooley and other social researchers envisioned the press as an important source of community values, the basis of social consensus, the "thread that holds society together."[4] Walter Lippmann stated the prevailing belief: "The power to determine each day what shall seem important and what shall be neglected is a power unlike any that has been exercised since the Pope lost his hold on the secular mind."[5] Based on such assessments of the power of the press, research on media influence was linked to the study of propaganda and its effect on public opinion and behavior.

In the late 1930s a growing body of research challenged the earlier assumptions, suggesting that the press was seldom the direct source of attitudinal or behavioral change. In light of this research, Paul Lazarsfeld and his colleagues developed a multistage model of communication effects in which information flows from the press to opinion leaders and through these leaders to the community. This model emphasized the mediating effect of personal relationships and primary groups such as the family in influencing people's response to media coverage.[6] However, by 1960 analysts were arguing that the United States had become a "mass society," characterized by atomization and the isolation of individuals. In this context, the relative influence of primary groups had declined, enhancing the direct effect of the media.[7]

An important book on the effects of mass communication by Joseph Klapper brought together these ideas. While acknowledging the influence of the press, Klapper argued that the images conveyed are assimilated and interpreted by different readers in different ways, depending on their prior beliefs, predispositions, personal experience, and the attitudes of their peers. In this view, press communication is but a contributing factor to, not the primary cause of, the public's attitudes and ideas.[8] Meanwhile, Marshall McLuhan and his followers extended the analysis of media impact beyond its influence on individuals by observing its effect on the political and cultural biases of the society.[9]

A common thread runs through much of the recent research; the effect press messages will have depends on the social context in which they are received. This context includes the educational background, personal experience, and reference group of readers. But alternative sources of information and imagery, such as television programs, comic strips, and other vehicles of popular culture are also part of the social context. These media convey a set of images about science and technology that are generally consistent with those in newspapers and popular magazines. In a content analysis of 1600 television programs broadcast between 1969 and 1979, George Gerbner, professor of communications at the University of Pennsylvania, found that science appears less in television news than it does in entertainment and science fiction programs.[10] These programs focus on situations of crisis and danger, and they usually portray scientists as forbidding and strange. While images of science as the source of modern day Frankensteins persist, the message is mainly that science is arcane.

Television documentaries present a different but related image. In an effort to personalize science, the scientist is made a star; the tweed and turtleneck chic of Carl Sagan and

Jonathan Miller represents a contrast with the eccentric and dangerous figures on entertainment programs, but these scientists are equally idealized. Many documentaries, such as those produced by NOVA, are thick with awe and reverence; while explaining science carefully, with elegant visual images, they, too, perpetuate the image of science as arcane.

George Basalla, professor at the University of Delaware, has studied the depiction of science in popular culture; comic strips, for example, portray scientists as "remote, superbly dedicated, logical and humorless individuals, apt to put reason above human considerations." The stereotypical scientist works in isolation, beyond public scrutiny. He is "a Faust-like figure who has great power over natural forces. . . . Because of his evil or damaged nature or because of circumstances beyond his control, the power of science is likely to be used against mankind."[11]

Marcel LaFollette, an MIT researcher, found readers exposed to a more benign but equally idealized set of images in American popular magazines between 1915 and 1955: "Mass media discussion of science combined extensive reporting on the actual results of science, promises and predictions that science would cure any social problem, and images of scientists as omniscient, powerful, well-meaning, and heroic, to develop a climate of expectations of what science could do for and to society."[12]

Similar expectations are conveyed by the ubiquitous "advisory arts." Readers of the extraordinarily popular "how-to" books and the syndicated health columns are given endless information, sometimes misinformation, about health and science. But above all they are presented with promises of future technological marvels and the image of science as an ultimate source of authority.[13]

The context in which images of science are received includes a large quantity of pseudoscience and superstition. The press itself not only runs a daily horoscope feature, but gives prominent attention to astrology and pseudoscience claims that purport to predict elections, natural disasters, economic trends, or personal affairs. Although such claims are usually downplayed or dismissed, their frequent appearance in the press lends them an aura of credibility and obscures the difference between science and superstition.[14]

For many readers, the only contact with science comes through articles in the tabloids so visible on supermarket shelves. The *National Enquirer*, the *Weekly World News*, and the *Star* all devote a lot of space to scientific and medical events or pseudoevents—mostly as a form of entertainment and titillation. Their stories take the press's preoccupation with scientific wonders and breakthroughs to its logical, if extravagant, conclusion. Scientists in the tabloids "play God" in their "secret laboratories." They can "cure splitting headaches," "create human beings," "guarantee boy babies," "save brain-damaged victims," and "cure impotence," "prevent cancer," and "find the key to long life."[15]

About 100 million Americans follow the comic strips analyzed by Basalla, millions read the tabloids, and most Americans watch television every day. Images appearing in these vehicles of popular culture help create the frame, the mind set, the expectations individuals bring to their reading of science news.

Today there is undoubtedly considerable public interest in science and technology. University of Michigan Survey Research Center surveys in 1979 and 1983 classified about 18 percent of adults in the United States as an "attentive public" for science and technology. These are people who, because of their educational and occupational background, are likely to follow scientific and technical issues. However, only about half of this group had a realistic understanding of the scientific approach, and many did not meet minimal criteria of scientific literacy. The survey data also suggested considerable and growing interest among the general public in science and technology. The percentage of respondents expressing interest in new scientific discoveries grew from 36 percent in 1979 to 48 percent in 1983; those interested in new technologies grew from 34 percent to 44 percent. The surveys, however, also indicate that this interest is not adequately met. Only 10 percent of the respondents in 1979 and 14 percent in 1983 felt very well informed about science and technology.[16]

Popular interest in science news often focuses on issues relating to health. Indeed, people use the press to learn about how to protect their health. A National Cancer Institute survey of how people become informed about ways to prevent cancer found that 63.6 percent get their information from magazines, 60 percent from newspapers, and 58.3 percent from television. Only 13 to 15 percent had talked to physicians about cancer prevention.[17]

The actual influence of the press, however, will vary with the selective interest and experience of readers. In esoteric areas of science and technology where readers have little direct information on preexisting knowledge to guide an independent evaluation (e.g., the effect of fluorocarbons on the ozone in the atmosphere), the press, as the major source of information, in effect defines the reality of the situation for them. During the period of maximum press coverage of the ozone controversy, for example, 73.5 percent of the general public had heard about this highly technical issue, previously remote from their experience, for the first time in the press.[18] From general studies of the effect of reporting, we can assume that where readers already have an established set of biases (e.g., about the cause of social behavior), science reporting tends to justify and reinforce these biases. And when the reader has had personal experience (e.g., with work place risks) or long-term exposure to press coverage (e.g., about environmental problems or dietary risks), the effect of media images is tempered by prior attitudes on the issues.[19]

What we know about public attitudes toward science and technology corresponds with the messages conveyed in the press. In 1965 historian Oscar Handlin described these attitudes, suggesting that science was hardly assimilated by the public even in the cultures of advanced industrial societies, which expend enormous resources to support the scientific enterprise: "Paradoxically, the bubbling retort, the sparkling wires and the mysterious dials are often regarded as a source of grave threat.... The machine which was a product of science was also magic, understandable only in terms of what it did, not of how it worked. Hence, the lack of comprehension or of control; hence also the mixture of dread and anticipation.[20]

Today such attitudes persist. Gilbert Omenn, a science policy analyst at the University of Washington, suggested that many youngsters perceive scientists as "geniuses or weirdos, people with whom they have little hope or no desire of identifying.... People close at hand are not recognized in their scientific roles. They don't fit the stereotype as brilliant problem solvers grappling heroically with the unknowns of nature...."[21] Yet surveys demonstrate a profound and unquestioning belief in the authority of science and the ability of scientists to solve problems and to make decisions in many areas of public policy.[22] Feelings of apathy, a sense of impotence, and a tendency to defer to expertise also are prevalent in the American population, as these surveys show. Their findings are certainly consonant with the media portrayal of science and technology as esoteric and arcane yet a source of authority and broad-ranging expertise.

Ironically, belief in the authority of science coexists with anxiety about its social effects—again paralleling the polarized imagery in the press. People tend to view science apart from its social context and to see it as removed from public concerns. Indeed, many people believe that the values of scientific research directly conflict with human values. A 1980 Harris poll found, for example, that 70 percent of Americans associated excessive concentration on science with "neglect of human problems," though 89 percent agreed that scientific progress is "necessary for a high standard of living."[23]

Public acceptance of science appears to be largely based on expectations of immediate applications. This emphasis on applications is often encouraged by scientists themselves in their quest for public support. It is perpetuated in the popular press through its emphasis on the immediate applications or implications of scientific work. And it is manipulated by corporate advertisers ("It is a medical fact . . ."), creating unrealistic public expectations about the benefits of science, and leaving the enterprise vulnerable when things somehow fail to "work."

Bad news about science and technology often affects consumer behavior, especially if alternative products are available. Thus, following the ozone controversy, people bought fewer aerosol sprays.[24] News coverage of toxic shock syndrome adversely affected the sales of some brands of tampons. After extensive media reports on dietary studies relating cholesterol-producing foods with heart disease, consumption of beef, eggs, and fatty milk products declined.[25] Similarly, reports on the risks of excessive sodium consumption was reflected in increased use of salt-free food products.[26]

Publicity about the possible adverse effects of birth control pills and IUDs also resulted in a significant decline in their use. This decline, between 1970 and 1975, correlated directly with the press coverage of events such as Senator Gaylord Nelson's congressional hearings on birth control methods, the studies linking the use of contraceptive pills to the risk of strokes, and the FDA warning to women over 40 about the increased risk of heart attacks. Each news disclosure brought sharp changes in behavior.[27]

Deliberate efforts to use the press to influence behavior, however, do not necessarily have the effect anticipated. Despite extensive news coverage of the Salk polio vaccine when it became available in the late 1950s, relatively few individuals agreed to be vaccinated at that time.[28] Similarly, media coverage of the 1964 surgeon general's report on smoking and cancer had little apparent direct effect on smoking habits in the short term.[29] Although people seek information from the press to guide even the most personal decisions (such as choice of birth control technique), they use such information mainly when it corresponds to their prior inclinations.

A more general effect of press coverage is to establish a framework of expectations, so that isolated events take on meaning as public issues.[30] For example, before the accident at Three Mile Island, a series of less critical accidents at other nuclear power plants appeared as isolated local events. The extensive press coverage of Three Mile Island alerted people to the structural context of such events and placed the issue of nuclear power plant safety on the public agenda. Similarly, before 1975 few Americans had even heard of dioxin or of a toxic waste disposal problem. By 1980, 64 percent of the public had heard of the issue and were worried about the disposal of industrial chemical wastes.[31] The press, in effect, made the problem visible and defined a "frame" or context in which related events could be interpreted. In this way press reports set the public agenda and ultimately influenced policy decisions.

The Framing of Public Policy

By their selection of newsworthy events, journalists identify pressing issues—the international competition in super-computer technology, for example, or the need of money for AIDS research. By their focus on controversial issues (e.g., toxic dumps), they stimulate demands for accountability, forcing policy-makers to justify themselves to a larger public.[32] By their use of images ("frontiers," "struggles") they help to create the judgmental biases that underlie public policy.

The importance of this last tool must not be underestimated. The real power of the press, claimed Harold Laski,
comes from "its ability to surround facts by an environment
of suggestion which, often half consciously, seeks its way
into the minds of the reader and forms his premises for
him."[33] Metaphors in science journalism cluster and reinforce one another, creating consistent, coherent, and therefore more powerful images which often have strategic policy
implications. When high technology is associated with
"frontiers" that are maintained through "battles" or "struggles," the imagery of war implies that the experts should not
be questioned, that new technologies must go forward, and
that limits are inappropriate. But if instead, the imagery suggests peril, crisis, or technology out of control (as in the case
of certain risks), then we seek ways to "rein in" the runaway
forces through increased government regulation and control. Calling the weakness of science education a "problem
of educational policy" implies the need for considered, long-
term policy intervention; defining it as a "national crisis"
implies the need for an urgent, if short-term, response. If
science is incredibly complex and arcane, and the scientist a
kind of magician or priest, this implies that the appropriate public attitude is one of reverence and awe. But if science is simply another interest group seeking its share of
public resources, this implies the need for critical public
evaluation.

The terms used to describe a problem, the sources cited,
the perceptions conveyed, point the finger of blame and
imply responsibility for remedial policies. For example, defining the relative contribution of personal life style and
environmental hazards to the incidence of cancer influences judgments about liability and claims for compensa-
tion. Defining the causes of the Three Mile Island accident
or the Challenger explosion and the relative significance of
human error and equipment failure has serious consequences in terms of who must pay. And placing the blame

for toxic waste disposal implies responsibility for cleaning up contaminated sites.

It was the newspaper publicity about Love Canal that forced the New York State Department of Health to issue a local health alert in the region, and this in turn attracted the national press. National coverage led to congressional interest, eventually bringing changes in national waste policy.[34] The deluge of press coverage about toxic wastes over many months made it necessary to establish blame and restore order, and thus forced the reorganization and restaffing of the Environmental Protection Agency.

We can trace the policy influence of science reporting in other areas as well. It was the media coverage of the controversy over recombinant DNA research that led the mayor of Cambridge, Massachusetts, to organize a citizens review board to evaluate the wisdom of building a laboratory in the city.[35] News publicity surrounding the debate over laetrile as a cancer cure forced the National Cancer Institute to test the drug on human cancer patients even though the lack of therapeutic effects on animals normally would have precluded human testing.[36]

By creating public issues out of events, the press can force regulatory agencies to action simply out of concern for their public image. Within four days of the initial press reports on the experiments finding bladder malignancy in rats exposed to high doses of cyclamates, the commissioner of the Food and Drug Administration put in place the cyclamate ban. This was perhaps a record for the rapidity of an administrative response to a technical study. Why such haste on the basis of limited evidence? "We were afraid of a leak," said the commissioner. He feared that public exposure in the press would politicize the FDA's activities and bring about legislative oversight.[37]

Press coverage can also influence the financial support given to research, a fact well understood by scientists and their institutions. In the 1940s the proliferation of cancer stories in the press helped to convince Congress to give research support to the National Cancer Institute. The dramatization of infantile paralysis in the press in the 1950s attracted millions of dollars to the support of research in this area. More recently, dramatic news stories on AIDS, especially when they involved well-known public figures such as Rock Hudson, have helped to generate public funds for AIDS research. And widely published reports of declining U.S. leadership in high technology have influenced legislative decisions to support costly megascience and macroengineering research centers in universities.

Corporate decision makers are also influenced by press coverage of science and technology, especially when issues lie outside their day-to-day sphere of competence. As their source of information, the press can lay the groundwork for establishing the credibility of funding requests. The wave of interferon publicity whetted the appetite of industry, affecting decisions to focus on interferon research. It also increased the availability of venture capital for new biotechnology firms. During the peak of media hype—between 1978 and 1981—the increase in corporate and government spending on interferon studies greatly outpaced the overall rate of growth in biomedical research, even though the peer review ratings in this area were lower than in other fields.[38] In politically popular research areas, the press plays an especially critical role in determining research support and, therefore, in shaping scientific priorities.

Publicity about problems in science may not only affect the allocation of financial resources, it may also bring external controls and turn the interest of scientists toward or away from specific areas of research. In 1973, when the *Washington Post* reported the National Institutes of Health's proposed guidelines on fetal research, right-to-lifers demonstrated at NIH offices, calling for a total ban on such research. Subsequent protest actions, including the 1974 arrest in Boston of physician and researcher Dr. Kenneth Edelin for murder of a fetus, were widely reported in the press, and fear of harassment discouraged many scientists from working in this field.[39] The widespread reporting of the XYY controversy in the late 1970s over studies examining the effect of having an extra Y chromosome on criminal behavior also brought harassment to the scientists involved and discouraged further work in this area.[40] Press coverage of fraud in science in the 1980s spurred science funding agencies and professional journals to reassess their policies and to establish safeguards that would obstruct further fraudulent practices and reaffirm scientific credibility. However, reports of the problems involving science and technology can also expand research. The news coverage of environmental problems helped to generate the public concern that brought research funding and scientific talent to such new fields as ecology and toxicology.[41]

The potential impact of the press on the public and on policy decisions is a matter of concern to both journalists and scientists. Both communities recognize a responsibility to inform the public accurately about science. And both see problems in the present patterns of communication between scientists and the public. But they approach the problem from their own professional perspectives, their own cultural frames. Thus, to understand science journalism, we must look at the interaction of these two communities, each with their own normative assumptions, social biases, and professional constraints.

6

THE CULTURE OF
SCIENCE JOURNALISM

A staff writer for a Pulitzer paper, the *Sunday World*, once described the philosophy of nineteenth-century science journalism in the following way: "Suppose it's Halley's Comet. Well, first, you have a half-page decoration showing the comet. . . . If you can work a pretty girl into the decoration, so much the better. If not, get some good nightmare idea like the inhabitants of Mars watching it pass. Then you want a quarter of a page of big type heads . . . and a two-column boxed freak containing a scientific opinion which nobody will understand, just to give it class."[1]

Today both the content and style of science news are more sober, reflecting the norms of objectivity guiding the profession, as well as the backgrounds and biases of reporters themselves. But contemporary science journalism still bears the weight of its tradition.[2] The early efforts to communicate science to the public defined a role for science journalists, created expectations about their relationship to the scientific community, and shaped the attitudes and norms of an emerging profession. Thus, to understand the present-day style of science journalism, we must first consider patterns and precedents established many years ago.

The Development of a Style

In the nineteenth century science and technology appeared in the press in both serious and sensational form. In the 1830s the *Athenaeum*, a London newspaper, published pages on the regular meetings of the Geological Society of London. Both European and American newspapers in the late nineteenth century published the lectures of such leading scientists as Thomas Huxley, Louis Agassiz, and Asa Gray, who were traveling around to popularize their work. In 1872 the *New York Tribune*, for example, published John Tyndall's physics lectures in a special edition that sold more than 50,000 copies.[3]

Most science journalism in the nineteenth century, however, consisted either of directly practical information about new farming techniques, the latest home remedies and the like, or wildly sensational stories. It was the heyday of science hoaxes. In 1835 the press reported that astronomer Sir John Herschel had observed batlike human beings on the moon. A story in 1844 reported a three-day crossing of the Atlantic in a balloon. The newspapers published reports of scientific oddities and weird claims worthy of today's *National Enquirer:* the earth was flattening out, a man with a stomach could live without eating, and a woman with no stomach could eat without digesting.[4]

Coverage later in the nineteenth century was influenced by both the dawning awareness of the power of science and technology and the deep ambivalence toward the industrial revolution. Science seemed increasingly fascinating but obscure, powerful but somewhat dangerous as well. Popular magazines began portraying science in nearly mystical terms as an awesome and remote activity performed by omniscient individuals. This, after all, was the century of Frankenstein and Dr. Jekyll and Mr. Hyde.

The role of science during World War I (in particular Germany's exploitation of chemical research to manufacture explosives), together with the postwar proliferation of consumer goods, increased the public's awareness of the social and economic power of science. In words that sound familiar today, the press expressed concern about America's role in international competition in light of the German monopoly on patents in chemistry. A 1917 article in the *Saturday Evening Post* asked, "What's the matter with American chemistry anyway?"[5]

As scientific research expanded after World War I, increased public interest in science was reflected in a growing popular science press. This press focused mainly on applications; science became a way to get things done. According to Frederick Lewis Allen, "The prestige of science at this time was colossal. The man in the street and the woman in the kitchen, confronted on every hand with new machines and devices which they owed to the laboratory, were ready to believe that science could accomplish almost anything."[6]

Ironically, enthusiasm about science in the opening decades of the twentieth century also increased antiscience tendencies: witness the revival of astrology and mysticism and the antievolution activities of religious fundamentalists, who saw Darwinism as a threat to their values. The press was a vehicle for such antiscience views as well. An important development, and a common theme in the press at this time, was the widening gap in knowledge between the scientific expert and the layman. In 1919 the *New York Times* published a series of editorials on the public's incomprehension of new developments in physics and the disturbing implications for democracy when important intellectual achievements are understood by only a handful of people. Morris Cohen, a friend of Einstein, whose theory of relativity had become a symbol of obscurity, posed the dilemma in a statement to the *Times:* "Free civilization means that everyone's reason is competent to explore the facts of nature for himself, but the recent development of science, involving ever greater mastery of complex techniques, means in effect a return to an artificial barrier between the uninitiated layman and the initiated expert."[7] This barrier, of course, separated journalists, as well as the general public. But, even as journalists were put off by the complexity of science, they were enamored of the progress it implied. They conveyed an image of science as an economic resource, an instrument of progress, a servant of technological needs.[8]

It was in this context that the newspaper magnate Edwin W. Scripps, founder of 30 newspapers and a syndicated press service, organized the Science Service in 1921. Science Service was the first syndicate for the distribution of news about science. Scripps believed that scientists were "so blamed wise and so packed full of knowledge . . . that they cannot comprehend why God has made nearly all the rest of mankind so infernally stupid." He believed that science was the basis of the democratic way of life. And above all he believed that, given the enormous social and technological changes of the period, science news would sell. With the help of a prominent zoologist, William E. Ritter, he engaged the cooperation of the National Academy of Sciences, the American Association for the Advancement of Science, and some leading journalists to translate science into "plain United States that the people can understand."[9]

Early in the formation of the science syndicate Scripps pondered a critical decision about its relation to scientific associations. Should the syndicate act as a press agent for the associations or as an independent news service? While hoping to avoid simply disseminating propaganda, he chose the former role. The syndicate was controlled by trustees from the most prominent science associations, and its editorial policies were dominated by the values of the scientific community.[10]

Presenting science in a marketable, that is, readable form, the syndicate sold its articles to over 100 newspapers during the 1920s, reaching more than seven million readers—one-fifth of the total circulation of the American press. It laid the foundation for contemporary science journalism, giving the profession both a purpose and a style.

The Science Service's statement of purpose emphasized "the importance of scientific research to the prosperity of the nation and as a guide to sound thinking and living."[11] Ritter stated the service's purpose more specifically; he hoped that it would gain financial support for science and encourage the "mental attitude of science" among newspaper readers.

But selling science to the public, the founders concluded, meant making some compromises. Although anxious to avoid the popular style of yellow journalism, the first editor of Science Service, chemist Edwin E. Slosson, believed science writers had to compete for readers by catering to prevailing popular taste. Diagnosing this taste, he concluded, "It is not the rule but the exception to the rule that attracts public attention. The public that we are trying to reach in the daily press is in the cultural stage when three-headed cows, Siamese twins, and bearded ladies draw the crowd to the sideshows, while the menagerie tent is soon vacated." That is why, he explained, science is usually reported in short paragraphs, ending in "-est." "The fastest or the slowest, the hottest or the coldest, the biggest or the smallest, and in any case, the newest thing in the world."[12]

In presenting science to his readers, Slosson emphasized human interest, drama, and even romance. One advertisement for the service announced that "Drama and romance are interwoven with wondrous facts, helpful facts"; another that "Drama lurks in every test tube." Science Service articles cast science as a new frontier and scientists as pioneers and discoverers. "The pure thrill of primal discovery comes only to the explorer who first crosses the crest of the mountain range that divides the unknown from the known."[13] Scientists were in addition described as industrious, persistent, and independent; all in all, they incorporated the most positive values of American culture.

In keeping with its statement of purpose, the Science Service portrayed science not only as the basis for technological development and economic progress, but also as a guide to correct thinking and appropriate behavior. For example, it gave prominent coverage to eugenics on the grounds that the masses must understand "that the fate of the nation depends on how they combine their chromosomes."[14]

Slosson soon became recognized as the most influential interpreter of science to the public. His service, shaped by both perceptions of public taste and the values and concerns of the scientific community, created a market for science news and a pattern for the emerging profession of science journalism.

The dozen or so science writers who began their careers in the thirties perpetuated the ideas set forth by the Science Service and adopted a similar style. The best known of these writers were not formally trained in science, but they conceived of themselves as missionaries, committed to science at a time when few people cared.

William Laurence, a science writer for the *New York Times*, for example, believed that science was "a way to rise beyond the disappointments of the real world."[15] Waldemer Kaempffert, also working for the *Times*, wrote eloquently of his "boundless faith in science, a faith justified by past achievement." He envisioned engineers who would provide us with "thousands of small towns with plenty of garden space, low rents, breathing space...."[16] Gobind Lal of the Hearst newspapers defined his role as a science writer in similar optimistic terms: "My job is to create a public taste for science. We must make science accessible to the people and for the people."[17] These writers became the gurus of science journalism when it burgeoned as a distinct specialty with the rapid development of science and technology in the years following World War II. They expressed a view of science and technology and developed a style that is reflected in the heroic images of science that mark science writing today.

Norms of Objectivity

Contemporary science reporting also reflects early efforts to adapt the norms of objectivity to the practice of journalism.[18] Journalists no longer believe that real objectivity is possible, but they are expected to approach the ideal of neutrality and unbiased reporting by balancing diverse points of view, by presenting all sides fairly, and by maintaining a clear distinction between news reporting and editorial opinion.

The Society of Professional Journalists (Sigma Delta Chi) has formalized the norms of objectivity in its code of ethics, which requires journalists to perform with "intelligence, objectivity, accuracy, and fairness" and to free themselves from all obligations, favors, or activities that could compromise their integrity.[19]

Accordingly, many individual newspapers place constraints on journalists to assure their neutrality. Reporters on the *Washington Post*, for example, must avoid involvement in politics or community affairs that could compromise or seem to compromise their ability to report with fairness.[20] A Sigma Delta Chi survey of 900 news executives found that half of the editors would not allow reporters to accept free trips under any circumstances.

In the field of science journalism these norms are particularly emphasized in the coverage of risk disputes and other controversies. Reporters try to maintain balance by quoting scientific sources representing opposing sides of a controversy, whether it be over toxic wastes, artificial sweeteners, or sociobiology. As a reporter who writes on controversies put it, "As long as you don't fall into the trap of presenting just one side, you're playing ball." But this approach gives readers little guidance about the scientific significance of different views.

Objectivity in the press is an American ideal; European newspapers are expected to have an explicit partisan view. In *Objectivity and the News*, Dan Schiller attributes our ideal of objectivity to three related factors.[21] In part it developed as a reaction against the excesses of yellow journalism in the nineteenth century. Objectivity was also encouraged when it became economically important with the organization of the Associated Press in 1848. This centralized news-gathering service had to sell its articles to newspaper clients with quite diverse political views; clearly, a special type of reporting was required. A third explanation can be found in the growing influence of the scientific attitude in the nineteenth century. Central to this attitude was the belief that facts, standing high above the distorting influence of interests and pressures, can and should be distinguished from values. The press in effect adopted the ideals of science, at a time when science was becoming broadly accepted as an apolitical basis of public policy, a model for rationality in public affairs.

The links between the ideals of science and the norms of journalism began to be forged during the 1830s. Prior to this time the openly partisan "party press" had dominated newspaper publishing. With the appearance of the "penny press" in the early 1830s the character of American newspapers began to change. Depending on commercial advertising rather than political patronage, the penny press asserted its independence from partisan views. It adopted norms of objectivity on the assumption that factual reporting would promote understanding and enhance democracy. In May 1835 James Gordon Bennett, editor of the *New York Herald*, promised: "We shall endeavor to record facts on every public and proper subject, stripped of verbiage and coloring." Later, in 1840, he expressed the purpose of journalism: "I feel myself in this land to be engaged in a great cause—the cause of truth, public faith and science against falsehood, fraud and ignorance."[22]

The editors of the penny press believed in an objective world that was only to be described. Science, with its assumed reverence for facts, was the guide to proper journalism. In a society strained by factional disputes, expressions of ideology and political views were to be avoided. A poem in the *Philadelphia Public Ledger* in 1839 stated the goal of the emerging newspapers:

> *We strike for right, and will not spare a blow,*
> *In bold defiance still our ensign show;*
> *All cliques, all sects, all parties we despise,*
> *Above all partial motives proudly rise;*
> *The laws our guide, the good of all our aim,*
> *We yield no principles for transient fame.*[23]

The norms of objectivity in journalism were reinforced throughout the nineteenth century as a means to avoid factionalism, encourage the values of pluralism, and promote a democratic process based on equal public access to "facts." An 1884 handbook for journalists, for example, states the imperative of separating facts and values in reporting and relates this imperative to American democratic values: "It is as harmful to mix the two in journalism as it is to combine church and state in government."[24]

There were cynics; as Mark Twain put it: "First you get your facts, then you can distort them as much as you like." However, by the turn of the century, belief in science as the embodiment of neutrality and rationality was firmly entrenched. Thus, in the first decades of the twentieth century, scientific values were penetrating many social and political institutions: witness the increased emphasis on technical expertise in government, the growth of realism in literature and art, and the political reforms of the progressive movement. In this context a scientific—that is, neutral—presentation of the facts was defined as the enlightened basis of a responsible press.

This spirit of objectivity converged with the idea that scientific knowledge prevailed over all other forms of knowledge to shape the conventions of the press. In 1915, for example, the *New Republic* proposed that journalism schools seek to create "a morale as disinterested and as interesting as that of scientists who are the reporters of natural phenomena. News gathering cannot perhaps be as accurate as chemical research, but it can be undertaken in the same spirit."[25]

Following World War I, political propaganda and the activities of the growing profession of public relations led to an awareness of how facts could be manipulated and inevitably to a certain skepticism about the reality of the "value-free" press. Indeed, faith in the objectivity of facts was soon considered naive. Newspapers responded by adopting various practices to prevent the manipulation of information, including the use of by-lines, the careful identification of news sources, and an increase in explicitly interpretive reports. However, objectivity remained an ideal, for it served the same purpose for journalists as it did for scientists, helping both professions maintain autonomy and independence from public control. Just as scientific objectivity enhanced the legitimacy of science, so it enhanced the legitimacy of journalists, allowing them access to sensitive information and justifying their protection under the First Amendment.[26]

The ideal of objectivity did not limit the seldom qualified enthusiasm in the press for scientific, medical, and technical advances. Immediately after World War II, and even during it, journalists wrote of the "promises" of peacetime applications of atomic energy, the "progress" in aviation, the "revolutionary developments" in pesticides, vaccines, and drugs. Above all, they hailed the "cosmic breakthroughs" of the space program.[27] In the temper of the times, discussions of values, of risk, of objectivity, were suspended; neither journalists nor for that matter scientists gave much consideration to the social implications or even the cost of science and technology, except, in the cold war context, when weapons were involved.

Not until the 1960s were doubts cast on the meaning of objectivity as a guide to journalism. Reflecting the political mistrust of the period, media analysts suspected that the "objective" press was simply in collusion with institutions of power; objectivity was viewed merely as a mystification and as a convenient myth. Journalists' efforts to maintain objectivity, according to sociologist Gaye Tuchman, were but a "strategic ritual"; subjective perceptions inevitably enter their writing, while the tone of objectivity allows reporters to avoid responsibility for their views.[28] Todd Gitlin argued that norms of objectivity only reinforce the "dominant hegemony"; by conveying an image of neutrality journalists in fact strengthen the existing legacy of private control over production.[29] Harvey Molotch and Marilyn Lester denied the possibility of objectivity; all news is manufactured, for there is no "world out there" to be objective about.[30]

Journalists themselves became more aware of how their own values entered their writing. As one experienced science writer told me: "A good news story is filtered through values. In the course of writing about science, I absorb an enormous amount of fact and information. I have to sort it out. My ideas about the role of the public and private sphere, my values concerning the environment, all help me to organize my material. We understand that. Scientists often fail to understand; yet they, too, carry baggage."[31]

Yet the idea that standards of scientific objectivity can be met by fair and balanced presentation of different points of view persists, evident, for example, in the reporting of risks. Ironically, this notion of objectivity is meaningless in the scientific community, where the values of "fairness," "balance," or "equal time" are not relevant to the understanding of nature, where standards of objectivity require, not balance, but empirical verification of opposing views. Thus while journalists' norms of objectivity were modeled on scientific method, their implementation in reporting scientific disputes is very often a source of irritation to the scientists involved.

Changing Professional Ideals

In the 1960s the expansion of an advocacy press with a critical and reformist ideology forced the mainline press to reconsider the conventions of journalism, and news articles became more interpretive, investigative, and adversarial in character. Science writing reflected these trends. Moreover, as scientists themselves entered as advocates on various sides of the growing number of environmental and energy disputes, the political dimensions of science became news and the role of scientists as sources of objective and neutral information came into question.

During this time a number of science journalists began to criticize the "missionary" role of their own profession. In 1966 Henry Pierce of the *Pittsburgh Post Gazette,* for example, called science journalists "a bunch of patsies prone to uncritical acceptance of anything we are told by our authorities—our authorities being doctors and scientists." He observed that other journalists maintained a more healthy skepticism towards news sources: "But we, bless us, go in with our bright baby-blue pencils poised, faithfully recording anything our scientists—gods—tell us. Never does it occur to us that these guys too may have motives that are less than noble."[32] John Lear, editor of the *Saturday Review,* wrote, "The spirit of untrammeled inquiry and skepticism required of journalism in other fields must become a standard in science writing."[33] And David Perlman chided his colleagues: "We are in the business to report on the activities in the house of science, not to protect it, just as political writers report on politics and politicians."[34]

Breaking away from the constraints of objectivity, some reporters during the 1970s captured the spirit of social criticism by sympathizing with the questions provoking environmental controversies and the growing concern about the social impact of science and technology. They tried to become critical investigators rather than passive conveyers of scientific and technical information.[35] They began, for example, to write more about the social implications of new technological developments. They increasingly reported disputes: over the antiballistic missile (ABM), the supersonic transport (SST), food additives, nuclear power, and environmental pollution. And they became increasingly sensitive to the implications of science and technology for such diverse areas as business, medicine, environment and energy, and even crime control.

More recently, in the 1980s, science writing, like other areas of journalism, has returned to a less critical and more promotional style—except in the face of disasters such as Bhopal or the Challenger accident. The images of heroes, breakthroughs, and frontiers that once again fill the news reflect the political rhetoric of the 1980s as well as the conventions that are the heritage of the science writing profession.

These sea changes in the reporting of science and technology can be seen even in the careers of individual journalists. Consider, for example, the career and writings of David Perlman of the *San Francisco Chronicle*. Perlman is one of the most experienced and prolific science journalists in the United States. A former president of the National Association of Science Writers and winner of several journalism awards, he is also one of the most highly respected writers within the profession. His career has spanned the post–World War II period of dramatic scientific and technological growth.

Perlman earned a degree in journalism in 1940 with no special focus in science. After serving in the army during World War II, he stayed on in Europe as Paris correspondent for the *New York Herald Tribune*. He came to the *Chronicle* in 1951 as a general assignment reporter and began to write mainly about environmental issues. He was attracted to science in 1958, when a friend gave him Fred Hoyle's *Nature of the Universe*, and began to specialize in this area of journalism. Science was local news in the Bay Area, with its concentration of research universities and affiliated laboratories. Just by focusing on the "local" research at Stanford, Berkeley, and the Lawrence Livermore Laboratory, Perlman could essentially cover the most important areas of contemporary science and the latest technological developments. But in addition, Perlman has traveled all over the world, covering professional society meetings, space shots, planetary probes, and scientific expeditions.

An examination of Perlman's articles in three critical years—1960, 1972, and 1982—reveals some important shifts in orientation toward science and technology that are characteristic of many reporters.[36] His writing in 1960 on science and technology expressed both his enthusiasm and awe as a new science writer and the general post-Sputnik optimism. Many articles were about advances in medical research. His language was extravagant. Medical researchers were detectives and wizards seeking "clues," "probing secret structures," "unlocking stubborn secrets," operating with "flashes of insight." They were also fighters "mobilized into assault teams attacking disease," and they needed public support.

Computers, military and space technology, research instrumentation, and especially nuclear power were described with similar enthusiasm as "super systems." The rocket was a "powerboost in a race to space"; a computer was "a damned machine that talks back with a flicker of ten red eyes." The peaceful atom was "the key to the future" in the "atomic horse race." Perlman wrote about the use of atomic energy for atomic flight, cheap fuel, monitoring smog, space exploration, deep sea mining, and farming fish.

The hazards of technology, especially radioactivity and atomic waste, later to become a critical theme, received some attention. Yet in an article on pesticides he welcomed the development of a new chemical pesticide as a boon and a breakthrough, with no hint that it could be hazardous, although it was derived from poison gas. The articles on hazards conveyed few of the concerns that were to dominate Perlman's writing a decade later. Rather, they are reassuring reports of agreements designed to protect the public, suggesting, for example, that "we are keeping tabs on atomic garbage."

Paralleling changes in the public and journalistic views of science and technology, the tone and style of Perlman's writing shifted rather remarkably by 1972. There were no reports of breakthroughs, far less drama, and many more qualifications. His articles on medical science—even those on organ transplantation—insisted on the limits of scientific claims. Perlman also wrote critically about the potential legal and ethical problems of genetic research, the politics of sickle-cell anemia research, and the controversies over acupuncture. The articles on advanced technology remained enthusiastic, but he reminded his readers of mounting public hostility to costly projects such as the further exploration of space. He wrote on the hazards from oil spills, food additives, pesticides, nuclear power, and air and water pollution, always emphasizing the controversies surrounding these issues. Describing the growing number of disputes among scientists about "troublesome technology," he observed the "tumultuous hazards" of scientists' involvement in public issues.

By 1982 Perlman was expressing the renewed technological optimism of this period. His language was more inflated; in writing of medical research he noted the "revolutionary" advances at the "frontier" of genetics and "promising" new diagnostic techniques. In covering biotechnology he employed an aggressive language: "fast breaking competition," "explosive growth," "revolution," "turmoil." In covering basic research he wrote about "cosmic mysteries," historic findings, adventures, and curiosities (a fish that likes to get hot). Finally, reflecting the expanded public interest in ethical and value questions, Perlman wrote extensively on human rights, ethics, and secrecy in science.

Perlman and other science journalists adapt their writing to the spirit of their times and use their personal instinct of what readers—and opinion leaders—will find interesting. As one reporter told me, "I consider myself the average general public. I think that if it is interesting to me, hopefully it will be interesting to the people I write for." But journalists' selection of newsworthy issues and style of writing about them are also products of their educational background and social biases.

Social Biases of Science Writers

The early generation of science journalists were seldom trained as scientists. Perlman, who describes himself as typical of science writers over 50, claims that he hated science in college (Columbia University) and took only the minimum requirements to graduate. "I started out, wearing a snap-brim fedora on the back of my head, covering fires; wore a trench coat in Europe as a correspondent for the *Herald Tribune,* and when I first started covering science, I truly thought that bilirubin was the name of the patient a doctor was describing when she talked about advances in treating hepatitis! . . . When Sputnik went up, I couldn't even have told you why it didn't fall right back down. Everything else I've learned on the job—out of enthusiasm, covering stories, reading a lot of background. . . ."[37]

Walter Sullivan, another veteran science reporter, joined the staff of the *New York Times* as a music critic. World War II turned him into a foreign correspondent, and then a series of trips to the Antarctic in 1956 transformed him into a science writer specializing in geology, astronomy, and physics. He and his professional peers developed familiarity with science mostly on the job through expeditions, press conferences, and occasional seminars. Some journalists took on temporary scientific jobs or internships, acting, in effect, as participant observers. One spent two weeks with a scientist tagging turtles in Florida; another spent three weeks on an oceanographic research vessel; another went on an expedition to the Galapagos Islands. Others have worked in various research laboratories, learning the language and methods of science. The younger science journalists have a more formal science background, often having majored in science as undergraduates.

Most journalists who cover science and technology, especially those working for small-town newspapers, write about science only part of the time. And even general reporters, when covering national security, crime, trends in education, budget priorities, or health, must often touch on some scientific or technical issues. These generalists often find the science beat confusing. Afraid of technical complexity, they are more apt to avoid substantive issues. And lacking both training and experience, they are less able to evaluate what they are told.

Science journalists themselves are divided as to the importance of formal training in science. On the pro side of this debate, William Stockton of the *New York Times* argues that science-trained journalists can be more critical about shoddy research methods and are less likely to take what they're told at face value than their untrained colleagues.[38] Reporters who know too little about science may not know how to find technical resources, what questions to ask about a technical matter, or what exactly to make of the answers. Preoccupied with reaching a basic understanding, they have little time or energy to interpret underlying issues.

Numerous cases support the need for greater methodological sophistication among journalists. For example, during a press conference in which a scientist described his cancer research, a probing and experienced science reporter found out that inbred rats were used in the experiments, an error that called into question the validity of the results. He refused to cover the story. A less experienced reporter, however, wrote it up without comment, having failed to understand that the methodological flaw undermined the significance of the findings.[39]

While agreeing that there is a need for greater technical sophistication, some journalists argue that too much science education can handicap the reporter. An experienced reporter without a background in science, but with a generally critical eye, can use journalistic skills to force scientists to explain things carefully. According to one journalist, "The generalist can often get away with asking stupid questions better than the science-trained journalists." He admitted his quiet appreciation when a less specialized colleague asked naive questions that helped to clarify a complex issue. Moreover, if a journalist knows too much about a technical subject, his writing may become overspecialized.

More important, journalists trained extensively in science may adopt the values of scientists and lose their ability
to be critical. Journalists worry about this: in interviews they
suggest that "the science writer who is basically a scientist
looks at things in terms of what it means to the development
of science, but the journalist looks at it in terms of how it is
going to affect people and their quality of life." According to
some reporters, "Science writers who are scientists have
more reverence for scientists than do other journalists. . . .
They sometimes draw back from tough questions because of
the very deep respect they have for fellow scientists." Others, however, contend that it is the least experienced reporter who is the most reverent, that science-trained reporters are "more jaded and therefore less vulnerable."[40]

Whatever the effect of their technical background, political and social biases clearly influence the attitudes of science journalists and their choice of newsworthy events. This
is most evident in observing what issues fail to appear in
articles on major aspects of science and technology. For example, with all the stories of chemical waste at Love Canal
and Times Beach, and the general coverage of chemical carcinogens, few journalists have written about the workers in
the chemical industry. As an intrusive neighbor, Hooker
Chemical Company was newsworthy; as an employer it was
not. Even the extensive coverage of the lawsuits generated
by workers exposed to asbestos focused more on the bank-
ruptcy manipulations by Johns Mansville Corporation than
on the health conditions of work.

Science writer Paul Brodeur speculates on the reasons
for the remarkable dearth of articles on this subject. "I sub-
mit that if a million people in the so-called middle or professional class were dying each decade of preventable occupational disease . . . there would long ago have been a hue
and cry for remedial action."[41] David Burnham, science
writer for the *New York Times*, explains why most reporters
avoid the subject of occupational health: "A class bias exists

and I have to calculate my stories somewhat to that reality. The upper-middle-class people who read the *Times* can shrug their shoulders and see cotton dust as a very distant, as a minor problem affecting some poor textile worker down south. But they can identify with a story of environmental carcinogens. Cancer terrifies everyone."[42] Likewise, Wade Roberts of the *Texas Observer* (an advocacy newspaper) confirms that "while the classes that own the newspapers are sometimes willing to take a strong stand on environmental health, occupational health just doesn't interest them."[43]

When journalists do cover occupational health issues they go to official sources rather than to workers. In 1976 a dramatic incident took place at the electric boat shipyard in Connecticut: it was announced that 1200 workers were found to have traces of asbestosis. This generated 50 stories in local newspapers, but only one reporter interviewed the afflicted asbestos workers themselves.[44] Karen Rothmeyer of the *Wall Street Journal* explains this bias bluntly: "Middle class journalists who are used to dealing with middle class officials won't get off their asses to make the difficult effort to find people on the other side. . . . Too many reporters wind up being Establishment stooges, not because they're uncaring people, but because they're middle class and don't want to struggle with speaking another language with different people."[45]

Social class biases are reinforced by political preferences or, even, the lack thereof. Journalists, contrary to the views of many critics, who see them as belonging to the liberal left, tend to be apolitical or middle of the road. Herbert Gans, a sociologist who has studied journalists, characterizes them as "members of the extreme center . . . suspicious of the oratory, skeptical of grand plans, committed to rational programs to solve problems."[46] The profession, devoted to the notion of objectivity, discourages those with strong political convictions. Politicized writers prefer to work for the alternative or advocacy press. Even journalists who cover numerous risk controversies for the daily papers seldom question the "system"; they prefer to conceptualize each problem separately in terms of aberrant individuals, bad companies, or ignorance.

Those journalists who specialize in science writing are particularly likely to avoid expressing a political view. Like many scientists, they dissociate science from politics. They rely extensively on scientific publications such as *Science, Nature,* and the *New England Journal of Medicine* and feel an obligation to report on the latest scientific findings as they appear in these sources. They define their role as explaining science more than analyzing its politics. When questioned about the politics of science, they will argue: "If I deal with the politics of an issue, then I stop being a science writer," or "I want good science, not moralizing."

The assumption that politics is inappropriate to science, however, can lead to conflicts of interest that undermine norms of journalistic neutrality. In 1979 the AP sold a series of Alton Blakeslee's articles on cancer research to 230 newspapers. The AP failed to note in the articles that Blakeslee was a paid consultant to the American Cancer Society and had originally written the series for them. In light of the heated politics of cancer research, Blakeslee and others questioned the ethics of this oversight. But the AP editor could not understand their concern: "That's like saying that God is political."[47]

The lack of political perspective also fosters uncritical trust in scientific sources of information. In reporting on biotechnology, for example, journalists rely on information provided by scientists involved in biotechnology firms, seldom considering the economic stakes in biogenetic research that inevitably shape their views.[48]

The reverential attitude of most science journalists toward their subject further reduces professional skepticism. David Perlman sums up the vision: "The science writers' words have captured the nobility to which the human mind can rise when it turns from violence and bigotry to an exploration of the infinite order of the universe. Political writers grow cynical, city editors grow old; only science writers stay young and endlessly excited and appreciative of elegance. For science by definition is young and exciting and elegant."[49]

Captivated by science and regarding scientists with awe, most science journalists write about their subject in glowing terms—in much the same way that sports reporters write about prominent sports stars. The veteran science reporters who have been on the beat since the initial space launches at Cape Canaveral are especially eloquent about science and technology. This influential group of writers, having spent long hours together waiting for space shots, developed a lasting camaraderie. During that dramatic period of more than a decade they gained an overwhelming sense of the excitement of science and technology; they felt that they were a part of the action.

Perlman, for example, describes "the excitement of witnessing scientific achievement firsthand; to watch the data on Jovian radiation pour in from Pioneer 10 in its flyby past Jupiter; to sit with Arthur Kornberg in his laboratory as he draws two ring-shaped helices in his synthetic viral DNA."[50] He is not alone. Science, to Victor McElheny, formerly of the *New York Times*, is "the Icarus." Describing his role as "personal witness to a golden age of scientific exploration," he writes: "To be present in the Lunar Receiving Laboratory outside Houston, Texas, when the first box of igneous rock and dust from the moon was opened; at the Jet Propulsion Laboratory in Pasadena when the first vertical strips of the first panoramic photograph from an automated Viking station on the planet Mars appears on the screen—a romantic observer could liken such opportunities to being a spectator at Barcelona when Ferdinand and Isabella stood to receive Christopher Columbus on his first return from the New World, or at Deptford when Queen Elizabeth came down to see Francis Drake, back home from circumnavigating the globe, and dubbed him knight."[51]

Science writers also find their subject endlessly entertaining. Their language of magic and wonder ("it boggles the mind") expresses their pleasure in describing science and technology. Many of them are amateur science buffs, endlessly curious about scientific issues. "I'd love to be a scientist," a journalist told me; "When I work on a story I get to sit at the feet of the most luminous minds in the U.S. It's more fun than anything I've ever done," said another. At the same time science writers often feel insecure when they interview scientists: "I feel naked without credentials."

This attitude of awe and admiration differentiates most science writers from political reporters, who are much more inclined to look critically at the events that they cover. However, some maintain a certain distance and skepticism. Ed Edelson of the *New York Daily News* declares (only partly tongue in cheek): "I refuse to believe that my purpose is to keep the public informed about the true nature of scientific research. If I truly did tell people about most scientific research, it would mean an end to public funding of that occupation."[52]

Most science reporters tend to behave rather like sports writers: they have chosen their topic out of love for it. They talk science constantly, even when off their beat. They are driven, according to one writer, by "inner directed ambition"; according to another, by "a sense of wonder and awe." While general reporters try to maintain a certain detachment from their subject by rotating beats, most science writers (like sports writers) remain with their specialty for years. Discussing their work, reporters say: "I cannot think of any job I would rather have"; "It is constant change, constant excitement"; "We're in it because we love science and it is a job in which you can keep learning more and more about science all the time"; "Science has made some wonderful advances; of course we view it as positive"; "We want to sell science."[53] While the more sophisticated writers make fun of the trite language of "breakthroughs," the background and biases of science journalists, as well as the precedents of their profession, perpetuate this style of reporting. Perhaps more important, their biases lead them to identify more closely with their subject and their sources than do journalists in many other fields. This identification, however, places them in an awkward position between two professions with quite different expectations. They strive to maintain the respect of their scientific sources and to satisfy the ideals of science, but they must, first and finally, meet the constraints of their own profession.

7

CONSTRAINTS OF THE
JOURNALISTIC TRADE

In the fall of 1980 Paul Jacobs, a science reporter for the *Los Angeles Times*, heard a rumor that Dr. Martin Cline, chief of oncology and hematology at the University of California at Los Angeles Medical School, had conducted a genetic engineering experiment on patients in Italy and Israel with a fatal blood disorder, thalassemia. Cline had earlier failed to gain approval for this research at UCLA. This promised to be a good story. Genetic engineering, especially when it involved human subjects, was intrinsically newsworthy. If in fact Cline's experiment, which required injecting genetic material into defective bone marrow, was successful, that would surely be of interest. But whatever the result, the ethical implications of bypassing the Institutional Review Boards, which evaluate proposed experiments, assured the story's appeal. Moreover, it was a local Los Angeles event, and no competitive paper seemed to be onto it. Thus Jacobs was able to convince the *Los Angeles Times* to give him the resources to investigate the story. His editors sent him to Israel and Italy "to flesh it out"; they gave him money, and they gave him time.

Jacobs first checked out the rumor, learned about the technical details of the issue, and obtained the names of people he could call. He then contacted scientists throughout the United States who might be familiar with Cline's work. Some scientists did not return his calls. Others reassured him that Cline's behavior was aberrant; they feared the incident would encourage regulations on experiments, constraining further research in the field. Jacobs then interviewed Cline himself.

In a hurry to avoid being scooped, Jacobs published his first story immediately after this interview. But then, having been promised a series, he continued to research the story and to develop its background and context. He interviewed geneticists at UCLA, research administrators, government officials, and Cline's associates in Italy and Israel. He obtained some documents by using the Freedom of Information Act, but most he was able to find through people he knew. At times the research became difficult, as Jacobs ran up against the self-protective instincts of the scientific community: "They don't want to give reporters or anyone else access to anything." He wrote up the research in several articles over a period of months. His stories all appeared on the front page.[1]

Lawrence Altman's long series of articles on AIDS in Africa in the *New York Times* in November and December of 1985 represented a similar kind of investigative reporting. Like Jacobs, Altman had months to gather and understand difficult material, the budget to travel in order to observe clinical settings and interview key people, and the space to explain technical material and medical events in their appropriate context. Both Jacobs and Altman worked under highly favorable professional conditions. But these conditions are relatively rare.

Most journalists working for the daily press are constrained by competition, deadlines, budgets, and the need to cover complex subjects within limited space and time. They must attract and hold the attention of readers, and they must develop an angle that will define their writing as news. These constraints, as much as the culture of science journalism, contribute to the character of contemporary reporting on science. Implemented through editorial policies and practices, they affect the work of journalists in all fields, but they pose special problems for those reporting on the complex, uncertain, and often slowly evolving events that characterize science news.

Newswork

Science journalists obtain material for their stories from press releases, public relations officers, professional society meetings, press conferences, scientific journals, and personal interviews. They usually write their stories very rapidly, pressing to meet their deadlines and to beat their competition on other newspapers.

Competition for priority affects the pace of daily newswork, encouraging a focus on "breaking news" and discouraging the coverage of long-term issues or issues that require extensive technical background.[2] "It's the constraint of getting out daily stories," reporters told me. "We can't take time off to do something that's not breaking news. We go with something that is there." Once the *Los Angeles Times* had scooped the Cline case, competitive papers were less interested than they might otherwise have been. The *New York*

Times, for example, ran the story, but in a short column on an inside page. It was no longer the latest development.

Emphasis on breaking news is often detrimental to good coverage of science, for important progress may not be associated with striking single events, and significance usually lies in long-term consequences. In the case of heart transplants or the implantations of the artificial heart, the significance of the technique rests on the patients' postoperative history. Yet media attention tends to wane after the initial dramatic event, unless, as in the case of William Schroeder, repeated strokes keep the issue alive.

Similarly, the focus on breaking news limits analysis of the methods and processes of science. Though important in assessing the significance of research, the methods of science are not considered news. Experienced science writers understand the importance of research methods; they often treat press conferences much like a graduate seminar, questioning research procedures and experimental controls. However, when they actually write up a piece of research they tend to focus on new findings or discrete events.

Because breaking news is usually scooped by the daily press, writers for weekly magazines are more inclined to provide background material and to write reports on scientific or technological trends. However, for most magazine editors science is a low priority. A *Newsweek* reporter identifies the "Friday Syndrome": "Every Monday, editors decide to do special pieces on science . . . by Friday, a disaster will take place or Congress does something controversial and our editors push these pieces off the seat."[3]

Science newswork is also constrained by limited budgets. Most papers under 100,000 circulation employ only general reporters, who cover several beats, because engaging a specialized reporter to cover science is so costly. Only the largest papers are likely to have separate specialists for, say, environment, medicine, energy, and science. Similarly, few papers allocate significant allowances to reporters on science for travel to meetings or interviews. While the *New York Times* provides generous budgets for leading writers like Lawrence Altman, and the *San Francisco Chronicle* has sent David Perlman on special assignments around the country and the world, most science reporters must compete with their colleagues in other fields, particularly politics, for travel money.

Science writers also compete with political writers for space and complain of problems convincing editors that science is as newsworthy. Journalist Earl Ubell once described the struggle of the science writer who has an ongoing, important story: "Frequently the first time he writes a story it appears somewhere behind the financial pages in the newspaper. The second time he writes the same story, with approximately the same lead, it moves forward so that it might get on the second page of the paper. And, finally, the third time around it almost always hits the front page because by that time the editor has begun to understand that perhaps it is news and that he ought to do something about it."[4] The amount of space allocated to science has increased over the past few decades, but with few exceptions, constraints on space limit the possibilities for including the background material and qualifications that are useful in conveying complex technical issues.

Deadlines present a further constraint. Shortage of time limits the number of sources that a reporter can use. According to a study by media analyst Sharon Dunwoody, half of the articles produced at a scientific meeting by reporters who had daily deadlines cited only a single source.[5] Those sources—mostly experienced public relations people—who can provide information efficiently in a form that is predictable and easy to turn into a story are the most likely to influence the shape of the news.

Time constraints have been eased to some extent by the use of computers. Reporters with access to data banks can gather information more rapidly, but the demands of daily story production still encourage the use of prepackaged information that has been organized specifically for the press. Reliance on such preconstituted accounts as press conferences, news releases, and computer-controlled information is, of course, no substitute for personal investigation, but it is a practical and therefore popular means of getting a job done under the pressure and routines of daily newswork.

Editorial Constraints

Editorial judgments mediate the work of journalists. Editors play a critical role in shaping the news; however, their influence appears more a result of incremental decision than of grand design.[6] It is the editor who decides what is published, how each article is cut, and where it will appear in the paper.

Journalists, unlike scientists, relinquish virtually all control over the final shape or presentation of their articles. When a journalist hands in a story the editor first decides whether or not it will be used, and in what form. If it is accepted, it is copy-edited, headlined, and positioned in the paper in the evening, all without consulting the writer, who usually sees the story only when it is published the next day.

Even when relying on the wire services, editors choose and edit stories to fit their judgments about how to maximize reader interest. With the exception of those few who were once science reporters themselves, most editors are trained in the liberal arts and are not very familiar with science. A study of editorial judgment found that editors use different criteria to evaluate science articles than either science writers or scientists.[7] Concerned with attracting readers, editors evaluate news stories primarily on the basis of their color and excitement, while science writers' evaluations are based more on accuracy and significance.

It is commonly believed that readers are less interested in analytical reports about science-related events than in how these events will affect them personally. Thus, one editor complains that some science writers on his staff are so preoccupied with scientific approval that they tend to lose sight of who their readers are, whereas he judges articles according to "what my mother wants to read, my kids want to read, my wife wants to read." Even David Perlman, an editor as well as a science writer, comments on the pressures of being on "the other side" and having to weigh science articles against news on "politics, murder, arson, fire, theft, corruption, war, revolutions, strikes, hot tubs and the higher California consciousness."[8]

Editorial control over the selection and presentation of articles is often a source of irritation among journalists. "When they misrepresent the story in a headline," reporters told me, "we give them hell, but we can't do anything about it." The placement of articles is another sore point. One reporter did a story on a large, important research center, only to find it placed on the back page when his paper did a front-page story on pickles: "Everything you want to know about pickles." There is, claims another, a sure way to make the front page. "Mention in the first paragraph something about a treatment for piles, ulcers or sexual impotence. Every editor has or worries about these conditions."

Another reporter complains that her editor sees science features as the "miracle page" and asks her to avoid writing so many dreary articles. A science writer for a popular magazine says her editor views science as either a "weird filler" or "a class act." He selects material that fits these categories and tries to include them in alternate issues. Editors, many writers claim, tend to change important qualifying words: for example, "may" is often changed to "is" in the interest of "style." Or they may inappropriately screen for technical terms: an editor changed a reference to Darvon from an "analgesic" to a "tranquilizer," for example, not appreciating that inaccuracy was the cost of using a more popular word.

It is often the editors who insist on definitive scientific

explanations even when there is real uncertainty. One refused to accept a report on Legionnaire's Disease that indicated that the scientists did not understand it. He queried
the reporter: "What do you mean they don't know? Get them
to explain what caused it."

Despite their frequent complaints, a number of science
writers say that they have greater autonomy than other journalists. Not only do they generate most of their own stories,
as opposed to receiving assignments, but they also enjoy
greater freedom to exercise personal judgment in writing
these stories: "It's a joy to cover the beat the way I think it
ought to be covered"; "I have more freedom as a science
writer than I ever did before." And their stories go more
lightly edited than those of political reporters. It is because
of the specialized nature of their subject that they enjoy such
autonomy: "Frankly, if I'm wrong, there's nobody here who
will detect that. . . . They just can't challenge me because
they don't know. A lot of responsibility is thrown upon myself as a writer."

But science writers often maintain this autonomy only
by exercising self-restraint. Journalists develop a symbiotic
relationship with their editors in which self-censorship limits direct editorial control. As one reporter puts it: "Every
time I write a story, I always have in mind if it is going to be
acceptable." The experienced reporter knows what will be
acceptable to an editor and will seldom push his own preferences to the point of confrontation.

Acceptable editorial policy, like journalism as a whole,
changes over time. During the Nixon administration the
adversarial relationship that developed between the Executive Office and the press resonated throughout the profession of journalism. The press demonstrated its capacity to
exercise enormous political influence: first the Pentagon
papers, then the Watergate affair became powerful media
events. For a time the press was less a means of publicity for
those in power than it was a power in its own right. Just as

political news was often critical of governmental policy, so too were reports on science and technology. In response, political administrations began to spend increasing amounts of money on information services and public relations to enhance their public image; so too did scientific and technical institutions. Media consultants have emerged as a new profession, significantly dampening adversarial tendencies in the press.[9]

Even so, many editors now fear that the power of the press will be turned against the institution to threaten its First Amendment protection. Indeed, legal and political decisions are eroding some of the freedoms long enjoyed by the press. Journalists lost about 85 percent of 106 major lawsuits between 1976 and 1983. Though many decisions were reversed on appeal, the threat and the high cost of litigation have had a chilling effect on investigative reporting. Even in areas not necessarily associated with lawsuits, such as science reporting, the fear of litigation, together with other factors discussed, has encouraged a carefully orchestrated type of journalism characterized by increased use of press releases, adaptation to prevailing (i.e., conservative) political values, and safe reporting that avoids sensitive issues.[10]

The tendency toward editorial caution is reinforced by evidence of declining public confidence in the press.[11] According to a survey by the National Opinion Research Center, in 1976 29 percent of the population had a great deal of confidence in the press; by 1983 public confidence declined to 13.7 percent. Other polls have indicated that a sizable number of people (20 percent, according to a *Los Angeles Times* survey) believe that press freedom should be regulated to avoid abuse. The restrictions imposed on the press coverage of the Grenada war, for example, received wide public approval.

In this context of declining public support, fear of legal action, and increased government pressure, editorial policy has become, in the words of an editor, "more judicious." Editors no longer "want to turn over a rock just to see what sort of dishonesty is under it."[12] And, partly because of self-restraint, many reporters, including science writers, share these cautious attitudes. They talk about "muzzling themselves," "conforming," "compromising their interests," "self-policing." Events that appear to run counter to prevailing values tend to be avoided. As a writer from a news magazine explains: "You have to be practical. What I want to do a strong story on is not always what my editors want a story on. We let a lot of things fall through."[13]

Audience Assumptions

Editorial constraints reflect perceptions of readers' preferences and ability to understand complex subjects. Seldom do journalists or their editors receive systematic feedback from readers. Yet, based on their personal observations, they maintain a set of assumptions about readers that heavily influence their choices and their styles.

Assumptions about audience preferences lie behind the many articles describing science in terms of its relevance for economic problems or its practical applications, even when these applications are far in the future. Recall, for example, the extensive coverage of biotechnology research, which focused for the most part on its influence on the stock market, or its role in international competition. Seeking greater relevance to maintain and increase readership and advertising revenues, science magazines in some cases have turned into service magazines, with numerous articles on computers and cameras. The editor of *Science Digest* ruefully explains why. He and other editors had assumed that in a high-tech world people would be interested in reading about science: "In hindsight that was a terrible assumption. For most people, science is some cerebral, lofty intellectual endeavor practiced by ivory-tower eggheads in white coats. For most people, it was a required course that ruined your grade-point average and made you feel stupid!"[14]

The effects of perceived audience constraints on the style of science writing are various. David Perlman, for example, likes to use colorful language in his leads: "If you warn readers they are about to read a science story, far too many will skip it as inevitably too complex."[15] While most journalists try to avoid a sensationalist and titillating style, they do tend to magnify events and to overestimate if not sensationalize their significance. Research applications, after all, make better copy than qualifications. "Revolutionary breakthroughs" are more exciting than "recent findings." Controversies are more newsworthy than routine events. Similarly, science becomes more palatable if made humorous or curious: (e.g., "Is Your Nose a Compass?" "How the Lizard Got Away," "Hungry Galaxy That Swallows Its Neighbors"). As one writer puts it, "We have to mix the pitches—to titillate, educate, inform, and entertain."

The focus on drama, aberration, and controversy in much of the reporting about science and technology reflects the quest of journalists to make their articles more entertaining. Medical stories, for example, are guided by what is popularly known in the profession as Cohn's First Law, named after Victor Cohn, a science writer on the staff of the *Washington Post*, who said, "There are only two kinds of medical reporting: New Hope and No Hope." News about technology and risk, as we have seen, is often conveyed with all the attributes of fiction, as a story with heroes and villains, conflicts and denouement.[16] Complex issues are usually avoided, for many journalists doubt that their readers will work sufficiently hard to understand them: "It's better to publish pap and get read than to starve." "Just try explaining the significance of DNA experiments to those more interested in the Giants' standing."

Finally, assumptions about audience influence the format of articles. Television has conditioned people to short bursts of entertainment and glossy information. During the 1970s many observers believed that the press would lose out in the competition with television, with its instantaneous communication of news and its visual impact. The press has survived, but it has become more like television in style. Some newspapers, for example, print shorter articles than formerly, give more attention to visual details, and employ language that is rich in visual imagery. The remarkable initial growth in the circulation of *U.S.A. Today* is often laid to its television format, with very short, profusely illustrated articles. Other newspapers as well try to attract readers through dramatic headlines, graphics, and leads. With detailed explanations and qualifications buried deep in the text, the images of science and technology received by casual readers who simply scan the headlines may be quite different from those received by careful readers.

Science writers explain their tendency to oversimplify by arguing that this is necessary to keep their readers' attention: "Those who scream about oversimplification are just not aware of the way people understand things." There are clearly numerous counterexamples: many articles on AIDS research, reports on nuclear reactor technology, and stories about the space shuttle accident have been detailed and complex, though directed to a broad lay readership. When an issue holds special interest, many people want technical details. Yet in an age where communication among scientists is specialized and obscure, simplification is an essential if controversial part of making science palatable to the public.

Economic Pressures

Behind the concern about audience interest lies the need for profit. Newspapers must maintain circulation, attract advertisers, and not offend their advertisers, their owners, or their boards. Newspapers are profit-making enterprises; most are owned by newspaper chains. In 1980 a mere 20 chains controlled half of the daily newspaper sales in the United States.[17] A study of the boards of directors of the 25 largest newspapers found that most members of the boards have ties to the corporations and universities that their papers cover regularly.[18]

Newspapers must operate according to the commercial realities imposed by their dependence on advertising. Advertising revenue, especially from local sources, has increased dramatically in the past few decades. The volume of local advertising in daily newspapers increased from $4.8 billion in 1970 to over $13 billion in 1980.[19] A typical newspaper derives about 80 percent of its income from advertising, which consumes about 65 percent of newspaper space.

The fate of science magazines and the science sections of newspapers rests more on computer, camera, and video ads than on subscriptions. It was a decline in advertising revenues, not a loss of readership, that led to the demise of several science magazines (*Science Digest* and *Science 86*) in 1986. Yet to sustain the necessary advertising level, the press is constantly struggling to attract readers—appropriate readers from a social class that would generate advertising income.[20]

Commercial constraints on what is published are seldom directly imposed; an editor, William Allen White once argued, "The publisher is not bought like a chattel. . . . But he takes the color of his social environment."[21] The editors and publishers of newspapers, especially in small towns, are often friends of corporate executives and may be constantly exposed to social pressure. One editor who joined a local country club described how the people he met there always wanted "to get something into a newspaper or to keep something out." For the reporter in this sensitive economic and social context, caution encourages routine, noncritical interpretations: "We are trained subtly not to cover the corporate sector and other private institutions the way we cover City Hall."[22]

The pressures on the policies of local newspapers are apparent if one compares the coverage of similar events in different areas with different economic stakes. Consider, for example, the reporting of environmental accidents in two communities. In 1969 the local press in Santa Barbara extensively covered the oil spill on the beaches there and was generally sympathetic to the environmental movement. The accident, it was feared, would affect tourism, a major local industry. Thus the goals of the environmental movement were compatible with local economic needs, and this was reflected in the tone of the coverage. In an environmental accident the same year in Utah, thousands of sheep were killed when nerve gas escaped from a Department of Defense project at the Dugway Proving Grounds. In this case the local press gave the accident very little coverage, suggesting its reluctance to challenge a project that was important to the local economy.[23]

Similarly, when 175 children in a northern Idaho town were found to have dangerous concentrations of lead in their blood, the region's newspapers shied away from the story. Pressure from the local lead and zinc smelter contributed to the news blackout. Moreover, scientists in the area were mostly employed by the industry and would not help reporters unravel the issue. Those stories that appeared had headlines minimizing the problem: "Doctors Said Lead Scare Out of Proportion," "Lead Poisoning Fears Largely Unwarranted," "Bunker Hill Warns Regulation Could Cause Shutdown."[24] Several media analysts refer to the "Afghanistan syndrome" in environmental reporting—the tendency to cover problems occurring in distant places or to think of problems as most serious in regions that are "up the road a piece."[25] The critical and ideological coverage of the Chernobyl accident is a recent case in point.

Some science reporters view these trends with cynicism but either restrain themselves or are restrained by their editors from bucking local pressures. Some typical comments: "When I got my first job in the newspaper, I saw all the Woodward and Bernstein movies and I read all the books on how moralistic and ethical newspapers were. The one thing I've learned is that newspapers . . . can be just as ruthless as any other company." "I cannot call a company a company without comment from the company, because we're so afraid of losing ads." "If you are in a one-industry town, to say that has no influence is naive."

The cost that can be incurred by defying economic constraints is illustrated by the experience of Richard Severo, a *New York Times* reporter who likes to raise contentious issues about corporate practices. In February 1980 he wrote a series of controversial articles on genetic screening of chemical workers.[26] The purpose of such screening is to protect the health of workers who may be genetically hypersusceptible to chemical risks, but critics of the procedure argue that companies can use it as an excuse to avoid remedying poor working conditions. In the course of reporting on the controversy, Severo questioned the implications of the procedure for the rights of workers. Specifically, he pointed out that Du Pont was only screening black workers, to find out if they carried the trait for sickle-cell anemia, although other ethnic groups carry traits for other genetic diseases. His articles thus implied that the use of such tests was discriminatory, transforming genetic information useful in family planning into a corporate tool.

Du Pont denied that the tests were used for job placement; the company was simply supplying a service for its employees. In a letter to the editor of the *Times*, Du Pont's chairman called Severo's series "sensational, erroneous, and offensive to management."[27] Severo then heard from a colleague that a Du Pont executive had boasted "We settled Severo's hash." Soon after, Severo claimed, his supervisor told him to "put the fire out," that the publishers wanted him to make less trouble. Severo, who had been a *Times* reporter for fourteen years and had won awards for distinguished science writing, was shifted to the metropolitan desk. The *Times* administration attributed the shift to Severo's publishing a book with an independent publisher, instead of Times Books as was then expected of staff reporters. Severo, however, interpreted the transfer as a punishment for covering politically embarrassing issues such as genetic screening.[28]

Severo's is an extreme story, unusual because external pressures are often less direct. But many science writers feel constrained by corporate pressure at the national as well as the local level: "I pay hell for doing consumer-related health issues." "Companies put pressure on my paper all the time. If I don't comply, I'd be punished."

Economic pressures are even more telling on television, which obtains its revenue from selling air time based on viewer ratings. Programming reflects efforts by the major networks to gain the widest possible audience and to avoid displeasing the corporate interests that provide advertising revenues. Thus the form and content of science information on television is conditioned by what producer Jeffrey Kirsch calls "the marriage of the sales mentality to the electronic image. . . . Producers have learned what formats, production, techniques, symbols, and personalities are most likely to succeed."[29]

Constraints of Complexity

The constraints of newswork, editorial supervision, audience assumptions, and economic pressures are faced by reporters in virtually every field, although the nuances may differ. Peculiar to science writing, however, is an additional constraint, that of having to assimilate and simplify vast amounts of sometimes extremely complex material. About

50,000 journals and about one million scientific papers are published every year, and rapid obsolescence of these papers has increased with the growing quantity of information. In certain active fields some 70 percent of the citations in scientific papers cover only the preceding five years.[30] This so-called information explosion has many consequences for journalists, who, even if scientifically trained, cannot possibly keep up with the latest details of science specialties.

Journalists covering a science issue may not know how to recognize what is important and may therefore miss a newsworthy story. They try to prevent this by keeping an eye on the more experienced science writers: "If you see Walter Sullivan twitching his withers and whinnying, then you know Walter is going to do a big piece and you know that your editors are going to see it in the *New York Times* tomorrow. You'd better write something on it or be very secure in your judgment that it's not worth it."

Few journalists or their readers can judge if numbers are meaningful or accurate, or if sampling techniques or research methods are appropriate. Furthermore, explanations in science often defy common sense. As a journalist puts it, "The news part is not hard to get right. . . . It is the mechanism by which something works that is hard to grasp and to simplify. How does the birth control pill work? How do you measure the rate at which the universe is expanding?" Yet complexity requires such explanations; readers will need background material if they are to understand the significance of scientific events.

The complexity of technical information compounds the usual problems of maintaining accuracy in reporting. Accuracy is extremely important within the value system of the press. Good journalists check their information with care. They sometimes help each other with explanations of difficult concepts or controversial issues, using each other as experts at press conferences. "We rely on each other's background and knowledge." "We keep each other appraised of what's going on and even sometimes help each other with a black sheet [carbon copy]." Yet scientists repeatedly complain about the inaccuracy of science reporting: "Whenever I read about something in my field, it is usually wrong."

Studies that have tried to identify the extent and possible causes of error have produced mixed results.[31] In 1970 scientists were asked to evaluate science stories, and 94.5 percent of them found the specific stories about their own work to be accurate, although when asked about the accuracy of science writing in general they were far more critical.[32] In 1974 scientists reported a mean number of 6.22 errors per science article, whereas the accuracy of general news has been estimated at a mean of about 1 error per article.[33] Other studies suggest that 40 to 50 percent of scientists complain about inaccuracy. However, when pressed, they usually identify the problem as one of omission of relevant information and lack of qualifying statements rather than of error per se.[34]

Often errors derive less from inaccurate reproduction of details than from the inevitable distortions that occur in translating complex technical terms into lay English. But they may also result from the writer's limited sense of the technical intricacies of an issue. For example, in covering the research linking saccharin to bladder cancer, many journalists failed to understand a standard scientific procedure based on extrapolations from large-dose experiments. Thus they interpreted the research in misleading ways.

The technical uncertainties characteristic of so many newsworthy issues in science compound the problem of complexity.[35] To report on health risks, for example, journalists must sort out uncertainties concerning the extent and significance of problems, their causes, the potential damage resulting from exposure, and the appropriate methods of remedial action—all of which may be subject to conflicting scientific interpretation. In many instances information simply is not available. Risk from toxic chemicals is a case in point. Relatively few of the 63,000 chemicals in commercial use today have been tested for their chronic health effect.[36] Compounding technical uncertainty is the difficulty of drawing precise causal associations between exposure to toxic substances and long-term health effects. Neither epidemiological nor laboratory studies can identify all the substances that may cause cancer, neurological problems, or genetic defects in humans.

The problem of associating cause and effect contributed to disputes over the effect of dioxin exposure on birth defects and spontaneous abortions at Love Canal. It confounded analysis of the health effects of food additives and the influence of ozone depletion on plant growth and human health. In these and other cases scientific uncertainty allowed a wide range of interpretation of risk. With few standards to guide their analysis, the specialists assessing risk arrived at different conclusions about the relative danger of particular hazards.

For journalists trying to write about these technical problems and to assess their significance, conflict among scientists presents a dilemma. Their difficulties are compounded by scientific jargon and the excessive information that is often provided to create the illusion of certainty and competence in confusing situations such as Love Canal. The quality of reporting in such cases varies, reflecting the journalists' technical sophistication and ability to sort out and interpret the available information. But, as we have seen, most often reporters simply try to present a balanced account by citing all sides of risk disputes without evaluating the quality of competing claims.

The complexity and uncertainties of the subject matter further reinforce the tendency of journalists to rely on press releases, press conferences, and other prepackaged sources of information. Journalists also attempt to compensate for the inadequacies of their own judgment by cultivating scientists who will give them a "scoop." But as historian Anthony

Smith suggests, "One man's scoop is frequently another's carefully camouflaged handout."[37] The difficulty in coming to grips with complex technical material reduces the likelihood of skeptical probing reportage, but even more important, complexity converges with the constraints of journalism to increase the science writer's vulnerability to those sources who try to shape the news.

Vulnerability to Sources

In May 1981, 175 newspapers around the country published a UPI wire service story about a scientist, Dr. Gregor, from a research institute called Metamorphosis, who discovered a wondrous all-purpose cure made from cockroach juice. The "scientist" had circulated press releases and announced his "findings" at a press conference. That reporters believed his hoax perhaps reflected their ignorance of Kafka and their respect for the invulnerability of the cockroach. But it is also a telling illustration of the vulnerability of many journalists when confronted with scientific and technical sources of information.[38]

Science journalists receive an enormous amount of mail. One writer told me that he scans 58 journals a month, receives about 250 press releases and 40 letters a week, and answers about 10 phone calls a day from scientists or their public relations officers who believe they have science news. He also attends about 20 scientific meetings a year and 2 press conferences a week. Another reporter said he finds several feet of press releases piled high on his desk every day.

Press releases are an American invention.[39] They first appeared in 1907, but only in recent years have they commonly been used as a means of shaping the news. Inundated with press releases, reporters are often cynical about them. Joseph Alsop once said, "All government handouts lie; some lie more than others."[40] A science writer for *Time* feels that they are "useless," "fit for nothing more than the garbage pail."[41] Another suggests that "every good reporter needs a good shit detector, and this holds true especially for science writers."[42] Still another finds press releases sad: "Sad for the eager publicists and their employers who undoubtedly labor hard to create the discarded press releases. Sad for the postal workers who must post this bundle of mail each day. Sad for those of us who must sift through the numerous letters. Above all, sad for the trees."[43]

Nevertheless, when time is short and information complex, reporters rely on press releases, often adopting their language as well as their content. In a study of sources of environmental information used by the media, David Sachsman found that during an eight-week period 11 environmental reporters received 1347 press releases. They relied heavily on these sources: over 50 percent of the published stories could be traced to these public relations efforts, and in many cases the reporters simply rewrote the release. Writing science news, Sachsman concluded, often involves simply "opening the mail."[44]

Science reporters also rely on press conferences at professional society meetings for information. These, too, they regard with some cynicism as "preselected and controlled news," or as "a source of free meals, free booze." They note that the American Cancer Society meeting is always timed to coincide with fund-raising campaigns. They complain of the "hype" about new cancer cures. "It won't be in clinical use for ten years and they give you the impression that it will be."

Although they feel manipulated, science writers admit their dependence on such preselected news. Thus most journalists covering a scientific meeting will report on the same stories. As one reporter put it, "In the presence of your fellow creatures . . . somehow it magically works out that everybody tends to agree on what is important." The less experienced a science writer, the more he or she is likely to rely on press conferences and to passively accept the material provided. More experienced writers use press conferences to meet people and to gain ideas, and then actively pursue stories and check them out through their personal contacts.

Who are these contacts? The problem of finding reliable sources and assessing where they stand in the world of knowledge is intrinsically difficult. The nature of science and technology encourages reliance on official sources of information—sources who are predictable and who know how to package information for the press. A study of the reporting about marijuana research found that the sources used by journalists were not the working scientists in the field, but rather the "big names"—the well-known people holding administrative positions."[45] Leon Sigal, documenting the sources of information used by the *New York Times* and the *Washington Post,* found that 46.5 percent were officials in government agencies, 4.1 percent were from state and local agencies, and 14 percent from nongovernmental groups.[46] A content analysis of the press coverage of the environmental controversy over PCBs in the Hudson River found that 43 percent of the sources cited were government bureaucrats. The *New York Times,* however, was more likely than local papers to cite scientists rather than public officials because the specialized reporters on the *Times* had many contacts in the scientific community and greater confidence in dealing with the technical information they provided.[47]

Experienced science writers know many scientists personally and have a "stable" of trusted experts, "inside dopesters" on whom to rely. But their sources do not necessarily represent the spectrum of opinion. A *New York Times* reporter told me in a somewhat embarrassed tone, "I prefer to interview those people I agree with the least. When I interview environmentalists, they offer insufficient hard data for me to publish their views. They may be right, but they have a conspiratorial attitude: there's a snake under every rock. They feel they are morally right and it's hard to get them to be specific. Industrial scientists, on the other hand, have data up to their eyeballs. They're smooth, they know what the media needs."

Less sophisticated reporters will often turn for expertise to those scientists who happen to be most accessible, for example, those teaching in the local university. Usually the reporter does not or cannot assess the experience of these scientists, their knowledge of the subject, or their reliability as sources. Faced with technical complexity and scientific claims that are difficult to check out, and socialized to regard science as the most reliable and objective source of information, these journalists are inclined to believe what they are told.

Experienced or not, science journalists are constrained by their concern about access to scientific sources. A *New York Times* journalist feels constrained less by editors than by the community she covers. "I've come from the ranks of science writers recently to join the political spectrum covering HEW for the *Times* in Washington, and I feel a lot freer covering politics than I ever did covering science. . . . It is very difficult [for science writers] not to be on the team. I am allowed to say things about the President, using my basic instincts as a journalist. That you just wouldn't think of doing in science."[48]

The vulnerability of science journalists converges with the economic and social pressures reflected in editorial policies and the other constraints of newswork to give an unusual degree of power to those sources who are best organized to provide technical information in a manageable and efficiently packaged form. Scientists and their institutions, increasingly motivated to enhance their influence over science and technology news, have become more sophisticated about how to do so. They exercise their influence primarily through two effective strategies: expanded public relations efforts, and increased communication controls over the dissemination of information to the press.

8

THE PUBLIC RELATIONS OF SCIENCE

In 1985 Kenneth Wilson, Nobel Prize winner in physics and an advocate of the development of supercomputers, helped to convince the National Science Foundation to provide $200 million for the creation of major supercomputer centers in four universities. He attributes the success of this quest to the publicity generated by a newspaper article. The article, on international competition in high technology, had cited a scientist's statement to the effect that without a nationally funded program the United States would lose its lead in supercomputer technology. "The most amazing thing to me," said Wilson, "was how the media picked it up and [how] a little insignificant group of words could change every-thing." This group of words had worked by establishing a powerful image. Wilson thus learned a basic principle in public relations: "The substance of it all [supercomputer re-search] is too complicated to get across—it's the image that is important. The image of this computer program as the key to our technological leadership is what drives the interplay be-tween people like ourselves and the media and forces a reac-tion from Congressmen."[1]

Other scientists and their institutions have come to simi-lar conclusions. The result has been the gradual develop-ment of a veritable public relations industry devoted to pro-moting science for the press.

Traditionally working in a context where success is mea-sured by the judgment of peers, scientists have long assumed that a record of accomplishment is sufficient to maintain research support.[2] Partly because of this assumption, infor-mation, the scientist's "stock-in-trade," has been directed pri-marily toward professional colleagues. Most scientists have not been interested in public visibility; on the contrary, they have feared it could result in external controls on their work. But attitudes in the scientific community are begin-ning to change. Increasingly dependent on corporate sup-port of research or direct congressional appropriations, many scientists now believe that scholarly communication is no longer sufficient to maintain their enterprise. They see gain-ing national visibility through the mass media as crucial to securing the financial support required to run major re-search facilities and to assuring favorable public policies toward science and technology.

This is not an entirely new perception among scientists; public relations in science goes back to the nineteenth-century development of professional societies. However, the scale of public relations efforts and the range of activities employed to attract the press have vastly increased in recent years.

Promoting Scientific Institutions

In 1847 Joseph Henry, the physicist and first secretary of the Smithsonian Institution, spoke of the need for greater public representation of science: "In carrying out the spirit of the plan, namely that of perfecting men in general by the operation of the Institution, it is evident that the principal means of diffusing knowledge must be the press."[3] In the nineteenth century, scientists such as Louis Agassiz, T. H. Huxley, and John Tyndall actively publicized their own work by giving popular lectures and writing directly for the press. However, with the increase of private philanthropy and industrial support at the turn of the century, scientists returned to their laboratories; popularization declined, and indeed, began to appear unseemly as a professional activity.[4]

After World War I an expanding scientific enterprise needed greater public funding, and once again scientists sought ways to enhance their public image. Professional associations began to organize public relations departments. In 1919 the American Chemical Society became the first scientific association to organize a news service. It hired a professional science writer to translate technical reports for the public and to write descriptions of scientific research for the press. In the 1930s a number of other associations followed the ACS lead.

Recognizing the importance of the press in creating a favorable public image, leading scientists were willing and eager to collaborate with Scripps when he founded the Science Service in 1930 (see Chapter 6). That decade also brought systematic public relations efforts from the medical profession. In 1937 the American Medical Association tried to establish rapport with science writers in order to counter the growing appeal of quack medicine. It created a press relations office run by a journalist, Lawrence Salton. Also during the 1930s, the American Society for the Control of Cancer used the press to implement a nationwide campaign to encourage early detection of cancer.

Public relations activities by scientists increased again in the years immediately following World War II. In 1952 the American Cancer Society (ACS) began the first of a series of laboratory and hospital tours for journalists, a veritable cancer marathon, to interest them in the goals of the ACS. The tour program, which continued for seven years, included visits to the universities, hospitals, and medical centers participating in ACS-sponsored research. One group of reporters visited centers in 21 cities in 20 days. The schedule prompted a poetic response, "The Saga of the Sarcoma Special":

> We heard of work on mouse and frog and newt and
>   toad and spider,
> And bee and fly and bird and chick and hen and Arthur
>   Snider.
> We heard of cortisone and glands, ACTH, anemia,
> Pituitaries, ovaries, adrenals and leukemia,
> Metastases, removals, series blocks and ugly rumors,
> The breast, the pancreas, the rays that never cure the
>   tumors,
> The palliative surgery which might destroy the bone,
> The seven danger signals which all seemed to be our
>   own,
> And when we'd heard it all from radiation to mitosis,
> We stood at last as experts in one field:
>   Tuberculosis[5]

As federal research funding expanded during the postwar period, interest among science institutions in popularization once again waned. But the 1957 Soviet launching of Sputnik brought renewed concern about the gulf between science and the public, and its implications for America's leading role in international affairs. Harvard physicist George Kistiakowsky expressed what many scientists felt at that time: the need for "the skillful interpreter who can translate scientific results and findings into language that the average reader can understand and appreciate."[6] Scientists once again discovered that, as a public relations officer put it, "the surest way to capture a share of the funds was to do good research and, almost as important, to talk about it."[7]

Scientists in the ensuing years supported the popularization of science out of ideological and cultural as well as economic concerns. In 1960 Jean Rostand, a French biologist and popular science writer, expressed a prevailing view: "The true and specific function of popularization is purely and simply to introduce the greatest number of people into the sovereign dignity of knowledge, to ensure that the great mass of people should receive something of that which is the glory of the human mind . . . to struggle against mental starvation and the resulting underdevelopment by providing every individual with a minimum ration of spiritual calories."[8]

Others, such as Jacob Bronowski, the British biologist and popularizer, talked of the need for a "democracy of the intellect." "We must not perish by the distance between people and power that destroyed Nineveh, and Alexandria, and Rome," Bronowski wrote. "The distance can be closed only if knowledge sits in the homes and heads of people and not up in the isolated seats of power."[9]

As the scientific and technical enterprise grew in complexity and importance, and the demands for research funds began to far outstrip the supply, scientists and research administrators increasingly emphasized the pragmatic goals of public communication about science. In 1971 a conference of biologists and health scientists, for example, concluded that the public must be given sufficient background material to understand the practical payoffs of seemingly irrelevant basic research so that it would be more willing to provide research funds.[10]

Since the 1960s and 1970s professional societies, academic institutions, and research organizations have all increased their public relations activities in order to enhance institutional prestige, encourage public support of research, and influence public policy toward science and technology. For example, the American Institute of Physics (AIP), which was founded in 1935, expanded its publicity programs in the 1960s, running seminars for journalists and news conferences to summarize newsworthy developments in physics. The AIP now routinely issues press releases and provides instruction to physicists on how to deal with reporters. The National Academy of Sciences, whose press office in the past primarily explained and interpreted technical reports for interested journalists, has recently changed the style of its relations with the press. According to both staff members and journalists, the academy has in the 1980s assumed a far more active role, initiating press releases and seeking maximum press coverage for their reports. In effect, it regards the press as a way to shape public attitudes and to influence congressional decisions on the funding of science.

Government agencies involved in costly technological or scientific developments play the same game. NASA, for example, developed a highly sophisticated public relations apparatus to attract media attention and win popular support for its costly program. So successful was the effort that it completely diverted press attention away from issues of safety and administrative mismanagement. It took the Challenger accident to shift the focus to such issues—and to the effect of NASA's public relations: "Some agencies have a public affairs office; NASA is a public affairs office that has an agency."[11]

Some scientific journals also have come to see the advantage of media publicity. Advance copies of the *New England*

Journal of Medicine and *Science* are sent by first-class mail to journalists, who must respect the mandated release date before writing stories on the articles. These competitive journals want to maintain their image as the key sources of scientific information for the public, and they skillfully use the press to this end.

Individual scientists have tried to attract press attention for a variety of reasons—to influence public views, to attract funds or to establish their competitive position in "hot" fields of research. In 1977 DNA researchers initiated a remarkable media campaign to show that genetic engineering research was safe, that its critics were irresponsible, and that regulation was unnecessary. More recently, leading computer scientists have been making extravagant claims to attract public support. Edward Feigenbaum, a well-known artificial intelligence expert, writes that with the fifth generation of computers "revolution, transformation, and salvation are all to be carried out."[12] Similarly, in his advocacy of the supercomputer, physicist Kenneth Wilson calls the new computer development "a second renaissance."[13]

Although individual scientists sometimes promote their own work, more often they rely on their institutions to disseminate information to the press. Most major research universities employ public relations professionals (called news officers or public information officers) or outside media consultants to publicize the work of their science faculty and, thereby, to enhance the image of their institution. Good public relations is important to these institutions, which must attract good students and staff, obtain money for research, and maintain public legitimacy. Their public relations professionals may be experienced science writers themselves, but unlike reporters, they work for those they write about. Their job is to insure that the institution's research is covered prominently, accurately, and favorably in the press. They make contact with journalists, set up press conferences, and write press releases. They try to make research appear newsworthy and timely, and they work with scientists to prepare them for media appearances. University public relations activities also include the preparation of brochures and news magazines describing the scientific research on the campus. These are mailed to alumni, science writers, and others. Stanford University, for example, which has one of the most active public relations offices in the country, mails the *Stanford Observer,* a newspaper largely devoted to describing Stanford's research, to 165,000 off-campus readers seven times a year.

Medical schools and hospitals also have extensive public relations activities directed toward influencing the media. They play to a receptive press eager to publish stories on artificial organ research, transplantation techniques, and reproductive technologies.[14] Favorable medical stories bring returns. When ten-month-old Jamie Fiske was admitted to the University of Minnesota Hospital to be considered for a liver transplant, no donor was in sight. Her parents went to the press with a dramatic plea. The case had all the elements of a good story: a personal tragedy, a family in despair, a father fighting bureaucratic obstacles to save his child, and the possibility of salvation through the wonders of medical science. The press responded, and news articles appeared throughout the country. They had an effect: a liver for Jamie Fiske, a letter from President Reagan, and money for both the family and research. Other transplant units enjoyed a remarkable increase in research funds as well as donated livers, including some offers from living people who, unfamiliar with anatomy, offered one of their own.[15]

At the Utah Medical Center, a team of public information officers worked with hundreds of journalists who were reporting on the first human implant of an artificial heart in 1982. The publicity paid off. The hospital received many donations of research funds, and Barney Clark's family received free housing and unsolicited gifts. Publicity had negative effects as well. Dr. DeVries, the surgeon who performed the transplant, received a series of bomb threats and a deluge of hate mail accusing him of interfering with God's will.

After Barney Clark died, the artificial heart team moved from the Utah Medical Center to Humana Hospital, a rapidly growing, for-profit medical care system which provided greater resources and fewer regulations. Humana promptly awarded a contract to a public relations firm to manage the media publicity and set up a media center staffed by seven professionals to handle press coverage of its first artificial heart experiment. The publicity surrounding Schroeder's artificial heart implantation in December 1984 cost Humana over $250,000. Humana's staff passed around briefing books with exhaustive information on the personal as well as the technical aspects of the Schroeder case. They provided telephones, places to work, and even food. It was, in the words of one reporter, "like covering a football game where they hand out statistics at the end of every quarter."[16]

The effect was especially striking if compared to the dearth of information available to the reporters who had covered the Baby Fae experiment only several months earlier. The implantation of a baboon heart in an infant in the Seventh-Day Adventist hospital in Loma Linda, California, had received extensive but vague and uninformed media coverage. The physicians on the case had refused to meet with journalists, who were left alone to search for leads and to speculate on day-to-day details. Indeed, they attended Seventh-Day Adventist church services to get information on Baby Fae.

Humana Hospital gained remarkable media exposure from the artificial heart, publicity that has subsequently served its reputation and helped to fill beds. But there, too, medical personnel soon turned off the information spigot.

After the initial burst of publicity, hospital physicians began to talk of the "media onslaught," describing it as "intrusive," "invasive," and an "ordeal." As Schroeder's condition worsened, Humana's approach to the press, according to a *Courier Journal* reporter, turned from what he called "NASA style" promotional publicity to "Soviet style" silence, which in effect tucked the experiment under wraps.[17]

The public relations activities of universities, medical centers, and other science institutions can, of course, be a useful source of information for journalists. Presenting complex material in a manageable form, public relations officers serve as liaisons between scientists and journalists, easing the job of reporting science. But what journalists gain in efficiency they may lose in reliability. Public relations officers know how the press works, and they use this knowledge to promote the interests of the institutions that employ them. This is a continuing source of concern to conscientious members of the news media. As reporter Robert C. Cowen of the *Christian Science Monitor* puts it: "We can deal with the Union of Concerned Scientists on one side of an issue and, knowing that bias, can deal with its other side. But when the bastions of professional purity and objectivity begin to worry about budget, jobs, and image to the point of reducing themselves to song and dance, where can science writers turn for objective background on scientific developments?"[18]

Some editors feel that their newspapers are used as pawns for grantsmanship. "When government money was available easily," an editor claims, "you couldn't get a story out of a molecular biologist. Today, I get copies of grant applications in the mail with this thing, 'single cure for blank,' or whatever the hell it might be, circled in red, saying 'we need all the help we can get, fellers'."

Individual scientists seeking publicity can themselves foster misleading coverage of their work. As we saw, scientists investigating gender differences in mathematical ability encouraged journalistic speculation on the possible implications of their findings for women's rights, sex roles, and job discrimination. Similarly, the scientists who were employed by and owned stock in biogenetic firms contributed significantly to the hype about the immediate benefits that would derive from interferon research.

Editors are especially concerned when the zeal of institutional or individual publicists results in the dissemination of misleading information. During the alarm over oil pollution in the Gulf of Mexico in August 1979, the news bureau of the University of Illinois told reporters that a professor had warned of a possible drought in the Midwest, based on a theory that the oil would cut off moisture from the gulf. The story, it turned out, was undocumented by meteorologists and simply reflected the promotional efforts of a university public information officer. Similarly, in April 1981 Harvard Medical School called a press conference to announce a new process transforming soft tissue cells in humans into bone. The press duly reported the process as a new discovery, only to find out later that the technique had in fact been developed through animal experimentation ten years before. The *New York Times* and the Associated Press then printed a correction of the story, announcing that it was, in fact, old news.[19]

In October 1984 public relations officers at the Dartmouth Medical Center held a press conference at which investigators announced the results of a preliminary feasibility trial of a therapy for Alzheimer's disease. The research was very preliminary, having been tried on only four patients, as was explained in a published technical paper. But the press release failed to mention the study's limitations, and the decision to hold a press conference turned the research into a media event. Not surprisingly, the press headlined the research as a "breakthrough," a "successful treatment," and a "possible cure," raising the hopes and expectations of readers. Twenty-six hundred people called the Dartmouth center to inquire about a cure.[20]

The quest for media visibility also prompted the University of Colorado press office to release a story about an archaeological expedition to an abandoned ancient city in Peru. The press release conveyed the impression that the expedition had found a "lost city," though in fact the ruins were listed in tour books, hardly lost except as a research site. Journalists featured the story as a discovery and only later realized their mistake.[21]

In these cases journalists, looking for a dramatic story and pressed for time, were inclined to believe their scientific sources and to rely on public relations professionals. But even when reporters suspect that publicity seeking lies behind dubious scientific claims, they may feel compelled to publish them. For example, all during the early fall of 1985 articles on the AIDS virus were appearing daily in the press, relating stories of awful illness and frustrating efforts by scientists in the United States and France to find cures. Then in November of that year a group of French scientists announced a "successful" trial of a drug that they claimed had reversed the course of the disease on several patients. Experienced science reporters were dubious. They knew that a trial on only a few patients was unlikely to be significant and that the time elapsed since the drug had been administered was too brief to mean very much. Moreover, in light of the intense international competition among AIDS researchers and the commercial potential of a new drug, they suspected the motivation of the announcement, which indeed was very premature. Some reporters preferred to ignore the story, but what were they to do? Given the public appetite for AIDS news, the preoccupation with the disease, and the competition in the news business, no scientific claims on this important subject could simply be ignored. Yet there was no way to check out the research. Reporters felt they had no choice but to report the information as provided by the researchers—and few had the confidence to report it with the skepticism it deserved.

Scientists in Industrial Public Relations

Just as academic institutions sell the importance of their science to attract a favorable press, so many corporations use the prestige of science to enhance their goals. There is, of course, a long tradition of using scientific images in advertising to enhance public confidence in products. But today firms are also using scientists themselves in their public relations efforts.[22]

Industrial firms have tended to develop and expand their public relations departments in response to crises that affect their reputation. Their formal public relations efforts developed at the turn of the century, first as an adjunct to advertising and later as a means to influence public policies that might bear on government regulation. Du Pont, for example, first formed its public relations department in 1934 after a Senate investigation of the gunpowder industry created an image of the company as "a merchant of death." More recently, after the Three Mile Island accident, the utility that ran the plant, Metropolitan Edison, expanded its public relations staff in order to counter its negative image in the press. Since the saccharin dispute, the Calorie Control Council has spent about $3 million a year on public relations for the artificial sweetener industry.

Industries engage scientists to provide technical information to the press, to enhance corporate credibility, and to legitimize company claims. Richard Tucker, a scientist and the president of Mobil Diversified Businesses, suggests why: "In an atmosphere like that of Times Beach or Love Canal or Three Mile Island, what has been generally missing is the voice of the calm, responsible scientist. To combat unreasonable fear, scientists must communicate more with the public and the media. . . . Public fear of chemistry is unlikely to be abated unless we try."[23]

One means of engaging scientists appears in a new kind of advertising published on the editorial pages of major newspapers. Called "advertorials," these columns resemble news items or editorials more than ads in their format, and they deal with contemporary issues such as government regulation or environmental health and safety. The chemical industry, for example, advertises its competence and public concern by printing photographs of scientists or engineers (sometimes accompanied by their children). Some of the captions are as follows: "My job is managing chemical industry wastes. What I do helps make the environment safer today and for generations to come"; "As a chemical industry engineer, I work hard to keep my community's air clean. After all, my grandchildren breathe it too"; "Just like you I want clean air and water"; "We will engineer out risks, impose detection techniques, expand studies"; "We have a technical staff of 10,000 specialists whose job it is to protect the environment."

Other activities of scientists in industrial public relations are more geared to shaping the news, especially the coverage of disputes over health risks posed by nuclear power plants, toxic waste disposal, and other products of modern technology. For example, during the controversies over the safety of nuclear power plants in the 1960s and 1970s, the nuclear industry developed an elaborate public relations apparatus that engaged scientists both at the national level, to convince the public of the safety of nuclear power, and at the local level, to show that utilities are good neighbors. In 1965 Hal Stroube, the public relations officer of Pacific Gas and Electric, outlined his strategies to "enlighten" the press. It was first necessary to gain the confidence of reporters, he said. "No one thing is more important to us. . . . We spent countless hours with the editors, science writers, and newsmen on dailies and weeklies around our 94,000 square-mile system. . . . We answered their questions honestly and fully. . . . We won the respect of the newspeople on our system and with it we won entry to the minds of our public through the news media."[24] As part of his strategy to influence the press Stroube proposed that industry spokesmen eliminate images and language that might work against them. He recommended that the Atomic Energy Commission (AEC) cancel a study on reactor accidents that could be used by antinuclear activists and that firms do "some semantic soul searching" to eliminate objectionable language: "palatable synonyms for scare words such as 'hazard' or 'criticality'" would facilitate public understanding of nuclear energy.[25] Thus nuclear plant sites became "nuclear parks" and accidents became "normal aberrations."

In 1975 consultants for the electric power industry outlined a "nuclear acceptance campaign," a strategy to "use the right medium to communicate the right message to the right target audience." After proposing appropriate messages to be directed to women, young people, and lower-income groups, the consultants advised that scientists should be the medium: "The public has faith in science, believes scientists and would listen." The campaign should "put articulate scientists out front."[26]

Accordingly, Westinghouse developed its Campus America program, designed "to increase positive media coverage of nuclear power." The company sent its own engineers and scientists around the country to lecture and make themselves available for interviews with reporters. Public relations firms that specialized in running political campaigns trained the scientists for public debate and taught them how to approach the media. From 1976 to 1982 the Westinghouse scientists and engineers visited 22 states and 125 university campuses, made 300 public appearances, and helped 300 newspaper interviews.

The chemical industry uses similar strategies. Michael Tabris, the director of corporate communications for Occidental Chemicals, sees the industry as "under siege" from a medium that intends "to lay bare the supposed transgressions of the powerful against the powerless." Occidental's strategy was to provide "the facts that rebutted the erroneous stories that were gradually becoming a part of the Love Canal background."[27]

Like their counterparts in the electric power industry, executives and public relations officers in the chemical industry are preoccupied with language and image. Industrial spokesmen say that the press is disseminating material that is creating irrational fear, hysteria, "cancerphobia," and even "chemophobia." They seek to use the press to teach the public that chemicals are natural, benign, and essential to life. Richard Tucker of Mobil emphasizes increased efforts to reach the press: "We must get across to the public the value of chemicals in our lives. . . . Scientific organizations and chemical companies alike must renew their efforts to find audiences to hear their story."[28]

In 1975 Dow Chemical established a "visible scientist program," sending scientists from the company on media tours to influence public opinion. Its goal in creating such a program was to affect the public response to the Toxic Substances Control Bill, then under congressional consideration. Later, when the press began to cover the link between cancer and some synthetic chemicals, the program was expanded. Each year Dow scientists ("credible scientists, not corporate spokesmen," according to company brochures) talk to civic groups and reporters around the country. Participating scientists have included toxicologists, biochemists, and environmental researchers. A public relations firm trains them in communications skills and on ways to project a rational, nonemotional image. The scientists travel to areas where chemical issues are of special public concern (for example, near major chemical plants or toxic waste sites). They convey the message that chemicals are necessary and that Dow is a leader in safety. In 1982 alone, 16 scientists visited 26 "media markets," held 24 newspaper interviews, and appeared on 62 television and 76 radio shows, reaching an estimated 9 million people with one or another variant of this message.[29] Often their arguments are reported as facts.

A New York City public relations firm, Hill and Knowlton, also runs a visible scientist program for corporate clients, arranging meetings between corporate scientists and the "right" editors. It has established an Industrial and Scientific Communications Service (ISCS), which publishes a monthly newsletter circulated to 2500 trade editors and science writers. Hill and Knowlton also prepares statements on controversial products for its clients. It orchestrated the Calorie Control Council's public relations campaign in opposition to the proposed saccharin ban. It was engaged by Metropolitan Edison after the Three Mile Island accident in order to increase its press credibility and by Ayerst Laboratories to offset the negative publicity about risks of estrogen replacement therapy.

Public relations efforts are also evident in the proliferation of conferences on science, technology, and the media, or on technological risks and the media. These conferences, often sponsored jointly by corporations and universities, bring journalists together with representatives of industry, government, and the scientific community. Their ostensible goal, as stated in the brochure of one such conference (convened to discuss the reporting of technical information on toxic substances), is "to provide a forum for open discussion of the various problems encountered by the media in obtaining and communicating accurate, well-balanced information on toxic substances to the general public. . . ." Industrial scientists and public relations officers, however, are the dominant voices at these meetings; seldom are neighborhood activists, union representatives, or other critics to be heard. Systematically focusing on the "problems" of the press in communicating technical information, these discussions are often simply a thinly veiled effort to create a science-based consensus that is compatible with the industrial agenda. By organizing the meetings, the sponsoring universities lend their credibility to promotional efforts.

Public relations officers have clearly observed that scientists have greater credibility than corporate clients. Thus, several public relations firms have advised corporations to develop "parachute teams" or "truth squads" of scientists, ready to move into risk situations in order to defuse the opposition by presenting technical "facts."[30]

The use of science in public relations is not limited to industrial interests. Environmental and science advocacy groups also use public relations techniques, engaging scientists to enhance their credibility and to legitimate their point of view. Members of the Union of Concerned Scientists tried to keep in touch regularly with reporters during the debate over nuclear power in the 1970s, for example. Individuals such as physical chemist John Gofman and physician Helen Caldicott used their credentials to win credibility for the antinuclear movement. Caldicott created a style attractive to the press by publicizing herself as both "scientist and mother." Antinuclear scientists worked with folk singers and rock artists to attract the media. Their books bore such colorful titles as *Poisoned Power* and *Nuclear Madness*. They drew Nobel laureates into the debate to publicize their position. In effect, they, too, saw the press as a means by which to create public attitudes sympathetic to their point of view while neutralizing the opposition.

Science is also a marketing resource. Corporations often try to sell their products directly through the press by publicizing new therapies as newsworthy scientific discoveries or significant medical advances. The publicity about estrogen replacement therapy, described in Chapter 3, was but one of many examples. In 1978 Merck sponsored a press conference on an arthritis therapy; this therapy was then hailed in the press as a "major medical advance," though it was primarily an effort to sell a product.[31]

Lilly's arthritis drug, Oraflex, was also marketed through science-based public relations directed at the press. In 1982 the firm's public relations office sent out 6500 press kits, promoting this new drug by making scientific claims of its effectiveness in relieving arthritis. Lilly also dispatched scientists around the country to contact smaller newspapers. Some experienced science reporters refused to cover the story, suspecting that Lilly's claims were exaggerated. However, the product was covered as science news by 150 newspapers and television stations. Prescriptions for Oraflex increased from 2000 to 55,000 a week. When a report showed its harmful side effects, the Food and Drug Administration (FDA) intervened and, after only twelve weeks, Oraflex was withdrawn from the market.

In 1986 the FDA again intervened to ask a drug firm to recall its press release on a drug called Virazole. The press release claimed that the drug's effectiveness in combating numerous diseases had been demonstrated by worldwide clinical research. Defining such publicity as an extension of drug labeling, the FDA accused the company of exaggerating its effectiveness and misrepresenting the safety profile of the drug in its communication to the press and thereby to the public.

Individuals also use their scientific credentials to market products. Several physicians have used science marketing techniques to promote the use of cortisol antagonists in the treatment of anorexia. This controversial therapy, designed to reduce the level of the hormone cortisol in the brain, was based on a study of 33 patients. Rather than submitting their work to a scientific journal, the physicians announced the findings at a press conference organized by a public relations firm and at a television talk show. "One can't afford to take the time it takes through the medical journals," said one of the doctors. The rush, according to an article in *Forbes*, was related to their effort to market a proprietary line of nutritional products and to expand a private anorexia clinic.[32]

In June 1984 David McCarron, a scientist with the Oregon Hypertension Program, reported in *Science* on research

indicating that lower levels of calcium in the diet are sometimes associated with high blood pressure or hypertension
and that calcium supplementation beneficially reduced blood
pressure. McCarron's research, which is partly supported by
the National Dairy Council (a trade research group for the
dairy industry), is controversial and a subject of debate in
the scientific literature. Other studies of the relationship between diet and hypertension point to the side effects of calcium supplements and argue that the data is too limited to
warrant dietary recommendations. Yet on the recommendation of the Dairy Council, McCarron hired two public relations firms to promote the *Science* article in the press.[33]

The boundaries between promotional news and commercial advertising are further blurred when prominent
scientists engage directly in commercial promotion. In 1976
Bloomingdale's advertised an "antiaging" cream developed
by the heart surgeon Dr. Christiaan Barnard with "a team of
Swiss cell biologists." In the ad Barnard writes about his
clinical experience with process of aging. "Unlike a defect
in the heart, the manifestation of aging is readily apparent
to everyone. . . . This has led my colleagues and I to identify
a substance much more abundant in younger skin than in
older skin. . . . The patented ingredients," he claims, represent an important "scientific breakthrough."

Public relations professionals see themselves as "an important arm of the media," a means to save editors hours of
work tracking down the news.[34] Their professional society
defines its principles as follows: "In serving the interests of
clients and employers, we dedicate ourselves to the goals
of better communication, understanding, and cooperation
among the diverse individual groups and institutions of
society."[35]

True, in the area of science, public relations officers do
contribute in important ways to informing the public about
products, ideas, and services. They often serve as "marriage
brokers," bringing together scientists and science writers,

teaching each group how to approach the other.[36] However, they also have to make their clients look good. Thus, many reporters see public relations less as a source of information than as "a means to promote, protect, and enhance the image of an institution, company, or product."[37]

From the earliest days of public relations, journalists have regarded such efforts as a means to subordinate journalism to private interests. In 1919 Frank Cobb of the *New York World* complained that direct channels of news were increasingly closed as information was filtered through publicity agents: "The great corporations have them, the banks have them, all the organizations of business and of social and political activity have them."[38] It was the influence of public relations on the news that prompted Upton Sinclair, in 1919, to define journalism as "a business in the practice of presenting the news of the day in the interest of economic privilege."[39]

Today, with stepped-up public relations pressure from science as well as business, journalists' negative attitudes have extended to this field as well. Science journalists complain of the endless stream of public relations professionals: "I get calls from Doctor Knowledge, the world's leading authority on X disease or Y technology who is also president of Z society." They refer to "pesky PR types" or "the flacks" who follow press releases with endless phone calls. "Any story that you might want to plant with me will, upon receipt of the phone call, have as much chance of making the pages of the *L. A. Times* as a run over dog."[40] One reporter describes the "law of public relations lunches: The quality of news you get is inversely proportional to the quality of the lunch."

Resentment of public relations is evident in the National Association of Science Writers (NASW), the professional society of science journalists. Reflecting the employment realities in the field, a growing percentage of NASW members are public relations writers working in universities or industrial firms. Though many are ex-science reporters and still freelance for science magazines, they are only allowed to be "associate members," unable to hold office or vote. Not surprisingly, they resent their "second-class citizenship" and joke about the "caste system" and "separate toilets." But debates about their status in the profession persist, as reporters continue to suspect their obligations to promote the views of their clients.

Journalists' suspicion of public relations is by no means limited to the information coming from industry: "They're all grinding the same axe, from breakthrough university to wonder pharmaceuticals to the National Institute of Nearly Cured Diseases." However, reporters still tend to trust scientists as sources, contrasting them to politicians: "When you talk to a scientist, you're talking to a fellow who is usually going to give you straight facts. His word is his bond. . . . When you talk to a politician, he is not worried about accuracy or truth; he usually reacts verbally and will say anything that comes to mind. He'll stretch the truth." Or, "We would rather talk to scientists than to politicians because we know that they are getting at the truth." The fact that journalists resent manipulation by public relations officers and overeager scientists does not diminish their influence— especially when sensitivity to manipulation is dulled by the prevailing faith in science as the ultimate, authoritative source of objective information. What is more difficult for journalists to accept is the reticence of scientists and their frequent efforts to avoid reporters. For a part of public relations is an increasing effort to withhold sensitive information or otherwise exercise communication controls over the news conveyed to the public.

9

HOW SCIENTISTS CONTROL
THE NEWS

One day in 1983 I talked to a geneticist who had agreed to hold a seminar on his recent research at the annual meeting of the National Association of Science Writers. He was worried and defensive. His scientific colleagues had warned him that he would be bombarded with value-laden questions about the potential applications of his work, its likely effects on industry-university relations, and its moral and ethical implications. In fact, the questions the science writers asked were limited to the technical issues he himself had raised; to his relief, the audience was simply struggling to understand the science. I asked him why he had been so concerned.

"Well," he replied, "we scientists are working in a competitive and precarious environment. We work in a conspiratorial world. Science," he said, "is a never-never land of extraordinary risk . . . publicity is suspect. Whenever we say anything in public we are concerned about our image. We're afraid of what our colleagues will say."[1]

Despite the generally friendly tone and positive, even promotional, images that characterize science and technology reporting, scientists complain of sensationalism and oversimplification. While they want their work to be covered in the press, they are constantly concerned about how it is covered, and this concern has led an increasing number of scientists and institutions not only to promote science through public relations, but to control journalists' access to information as well.

Blaming the Messenger

Seeking a press that expresses their views and supports their goals, scientists feel betrayed when their views are challenged or distorted. They defensively interpret critical reports about science or technology as evidence of an antiscience or antiestablishment bias. In 1980, for example, Philip Handler, then president of the National Academy of Science, wrote with reference to the coverage of environmental disputes that "antiscience attitudes perniciously infiltrate the news media."[2] George A. Keyworth II, President Reagan's former science adviser, asserts that reporters who cover science and technology deliberately distort facts. Accusing the media of irresponsibility, of emphasizing only the hazards and not the benefits of emerging technologies, he concludes that "the press is trying to tear down America."[3]

A scientist from UCLA wrote in a Dow Chemical publication, "The press often seems intent on showing how modern technology, including the chemical industry and nuclear power plants, is poisoning America." A chemical industry spokesman picked up on the image. "If there is any poisoning of America going on, it is not chemicals that are the culprit . . . it is the media, which all too often seem intent on burying us in piles of purple prose—a sort of verbal poison. In the process, journalists have helped to create crises where none exists (the cancer epidemic), have blown out of proportion legitimate stories (Three Mile Island) and avidly hunted for crises to come (acid rain). Now they seem bent on doing all this with the waste chemical issue."[4]

Even the most tempered and factual reporting can provoke a defensive response. When Harold Schmeck of the *New York Times* wrote the nonsensational article about interferon research noted in Chapter 1 in which he warned his readers not to hope for immediate miracles, a group of scientists complained that such expressions of doubt in the press would affect their research funding: "It is the public's will and support that make further progress possible."[5]

Scientists also complain about inaccuracy in the reporting of science and technology. When questioned further, though, they admit that inaccuracy is mainly a problem, not of getting the facts wrong, but of omitting qualifiers or details necessary to place information in a proper perspective. The omission of information on the methodology of animal tests in many accounts of the saccharin controversy described in Chapter 4 is a case in point. The nature of contemporary science journalism—the constraints of space and of time, the pressures for simplification, even the quasireligious faith in the capacity of science to provide definitive solutions, and sometimes the ignorance of the reporter—are usually to blame for such omissions. However, scientists often suggest an explanation that ignores these complexities: that journalists simply care little about the truth. Arnold Relman, editor of the *New England Journal of Medicine* (*NEJM*), puts this bluntly: "The press, the media in general, are much more interested in the story, the news, than in the facts."[6]

Whatever the reasons for such omissions and inaccuracies, many scientists and public officials have voiced concern about the possible harm from media coverage of scientific and technological events. After the Challenger explosion, for example, a NASA official declared that it was the press that had pressured the agency to jeopardize flight safety in the space shuttle program.[7] In the aftermath of the artificial heart implantation at Humana Hospital, leading medical scientists suggested that the extensive reporting of the events could be detrimental to medical science and was less than a public service. Joseph Boyle, president of the AMA, referred to the publicity as a "Roman Circus" in which the desire of "making the deadline for evening news" could influence medical decisions. Arnold Relman argued that the publicity was simply a form of self-promotion orchestrated by the for-profit hospital. In the interest of scientific objectivity, both advocated less public reporting until the results were quietly and dispassionately reviewed by the scientific community.[8] Following the initial publicity, even Humana's physicians, it may be recalled, became irritated with the press. Concerned about disruption (having found journalists hiding in laundry carts to get forbidden photographs) and reluctant to reveal the growing differences of opinion about the wisdom of the program, the hospital administration dismantled the media center and stopped the daily press briefings.

Occasional glaring incidents contribute to the sensitivity of scientists toward the press. Scientists often tell horror stories of being misquoted, misinterpreted, or even maligned. For example, the chairman of the International Task Force on World Health and Manpower relates his experience at a press conference. A reporter asked him whether he thought witch doctors can effectively administer medication in Africa. He replied that they probably could because of their high credibility in the population. The headline the following day read "U.N. Expert Calls for More Witch Doctors."[9]

However, the ambivalence so many scientists feel toward the press stems less from their personal experience than from certain characteristics of their own profession in the context of recent social and economic pressures. Part of the tradition of science is a strong belief in the value of open communication. Although secrecy has also been part of the competitive culture of scientists—often engaged, after all, in priority disputes—the sharing of data has long been considered both a moral imperative and a pragmatic need. For secrecy is believed to be damaging to science, an obstacle to creativity, to the cumulative work necessary for progress, and to the peer review necessary to maintain the quality and integrity of scientific work.[10] Furthermore, scientists have often taken action to protect the public's right to know by opposing security restrictions, loyalty oaths, trade secrecy, and other measures that would restrict communication.

In recent years, however, the scientific enterprise has been undergoing significant changes that are affecting scientists' norms of communication, and in particular their concerns about the press.[11] Science today has come to depend on sophisticated technology that is both extremely expensive and more susceptible to regulation. The increasing costs of science are changing the nature of both the scientific profession and its relation to the public, as criteria of social or commercial merit, external to science, are brought to bear on the funding and control of research. Increasingly, support for large-scale projects is sought from industry or directly from congressional appropriations, bypassing traditional agency sources (e.g., the National Science Foundation), which rely on scientific peer review. As competition for these corporate and government funds increases, many scientists are convinced that traditional professional standards are not enough, that their research support—and their ability to maintain autonomy—will be affected by their public image.[12]

Changes in the relation of science to the public also reflect the growing economic, social, and policy importance of the knowledge generated by research. Control over scientific knowledge is integrally linked to power over public affairs. Thus people are increasingly demanding technical information, especially in controversial policy areas. For example, growing concern about the health effects of toxic wastes, environmental carcinogens, and chemicals in the workplace has been reflected in "right to know" legislation, in the demands by citizens for research data, and in greater use of the Freedom of Information Act. And as scientific information is perceived as a political resource, scientists themselves become engaged in public disputes. Their widely publicized participation inevitably corrodes the public image of their neutrality and disinterestedness, and brings public pressure to bear on the process of science itself.

In general, while scientists see public communication of scientific information as necessary and desirable, they are also extremely aware that it extends their accountability beyond the scientific community. For once information enters the arena of public discourse it becomes a visible public affair, and the way is opened to external investigation and regulation. Increasingly vulnerable to such external pressures and concerned about threats to their professional sovereignty, scientists have begun to seek more control of the discourse on science—to influence the images of science in the press.

Strategies of Control

"You science writers live off the crumbs from our table."
"Unfortunately, sir, it is a hard life because the crumbs are so often stale."[13]

The scientific community has a well-articulated set of professional norms that govern relations among scientists.

However, unlike physicians and other licensed professionals with codes of ethics and standards of confidentiality, they share few norms to guide their relations with the public. Defining their work as an autonomous enterprise, scientists are ill-equipped to deal with the external pressures represented by the press.

In an effort to mitigate the problem, a number of scientific journals have published guidelines on how to respond to journalists and how to avoid misrepresentation. These guidelines are usually defensive in nature. For example, an article in the *NEJM* suggests that scientists use the public relations office of their universities as a clearinghouse. It warns scientists to be aware of reporters' motives: "Your response to an innocent-sounding question about your study of schizophrenia may be linked in tomorrow's newspaper to a murder trial." The article also warns *Journal* readers to avoid interviews about their research prior to publication and to be extremely cautious in what they say: "There are many instances in which a researcher has been led innocently to the slaughter," and "Never even whisper to a reporter anything you would not care to see in screaming headlines." It suggests that scientists who are to be interviewed first do a dry run with a public relations officer and also that they tape the interview so as to have an exact record. "If you feel trapped, obfuscate: it will get cut if it is too technical."[14]

Scientists attempt to control science news in part by discouraging their colleagues from "going public." In her book *The Visible Scientists*, Rae Goodell observes that scientists who become visible to the media "are typically outsiders, sometimes even outcasts among established scientists, . . . seen by their colleagues almost as a pollution in the scientific community—sometimes irritating, sometimes hazardous. [They] are breaking old rules of protocol in the scientific profession, questioning the old ethic, defying the old standards of conduct."[15] Those who have the confidence to violate these norms are usually scientists with academic tenure and established reputations.[16]

Scientists' inclination to avoid reporters is encouraged by the editorial policies of a number of professional journals: the *Physical Review Letters*, the *Archives of General Psychiatry*, the *NEJM*, *Science*, and several other journals will not consider an article whose content has been published in the popular press. These journals are highly prestigious; publishing in them can affect a scientist's career. Thus their policies serve as effective constraints on scientific communication—often in research areas that are of broad public interest. Such policies have thus been a source of continuing and acerbic discussion, which has focused on the "Ingelfinger rule" guiding the publication policies of the *New England Journal of Medicine*.

In 1968 Franz J. Ingelfinger, the late editor of the *NEJM*, decided that he would not publish a scientific article if the details of the article and its supporting data had been previously reported in another journal or in the press. He believed that the *NEJM* was more than simply archival; it should also be newsworthy. "The original article has an appeal quite different from that of the comprehensive survey. . . . The reader is more involved, his appetite is less dulled by the fact that his cerebral exposure to the news is direct, not through the dialyzing membrane."[17]

Ingelfinger's successor, Arnold Relman, has perpetuated the rule. While Ingelfinger developed the policy because he wanted the *Journal* to be newsworthy, Relman emphasizes his responsibility to maintain the reliability of scientific information through the system of peer review. He argues that the public interest is not well served by disregarding this system, for journalists could raise hopes or fears on the basis of false or unreliable information. In addition, Relman argues that prior disclosure places a burden on physicians, who should have the opportunity to read about research in an authoritative source before being besieged by patients clutching a newspaper article.[18]

Problems of implementation immediately occur because of the normal practice of prepublication presentation of research at scientific meetings. These are often well attended by journalists. Relman concedes that press coverage of papers presented at meetings cannot preclude publication. However, he advised scientists not to grant interviews on the details of their conference papers since their work at this stage is usually preliminary and incomplete, and may never warrant scientific publication.

Journalists are appalled by the Ingelfinger rule, arguing that it violates the public's right to know. They cite areas in which delayed journal publication as a result of press coverage postponed appropriate public health measures. For example, a newspaper story about early laboratory research on the effects of "smoking" on beagles prevented publication of the findings of this cancer research because scientific journals refused to publish the results on the grounds of prior disclosure in the press. Indeed, due to such concerns, some journals, such as the *Journal of the American Medical Association* (*JAMA*) have refused to adopt the rule. Relman himself has made exceptions in the *NEJM*; early findings of research on toxic shock syndrome and AIDS, for example, can be disclosed without jeopardizing the chance of publication because early public knowledge of medical information on these issues is urgent for decisions affecting public health.

Journalists, however, find that the policy significantly reinforces scientists' reluctance to talk to the press. Ironically, the very newsworthiness of *NEJM* (which, after all, is mailed to reporters before it is sent to subscribers) has created a struggle over the control of scientific communication.

Often scientists try to control press coverage by refusing interviews unless they can review and correct the copy prior to publication. Reporters, fearing censorship by vested interests, are usually reluctant to show their articles to sources, though they often confirm the accuracy of details with them. Science writer Victor Cohn suggests that "scientists are to reporters what rats are to scientists. Would scientists allow their subjects to check the interpretation of their behavior?"[19] Another writer, Earl Ubell, expanded on this characterization at a meeting of medical researchers: "As a reporter, I am neither proscience nor antiscience. To me as a reporter, the scientist is just like a rat is to you. I am looking at you through my microscope and trying to describe you. . . . What scientists and doctors do, not what you say about what you do, is ultimately what ought to be getting reported."[20]

Beyond their personal caution scientists may find themselves constrained in their public communication by corporate or government employers, or by government restrictions on information dissemination. Industrial employers, who are preoccupied with trade secrecy as well as their image, often place severe restrictions on the publication and communication practices of the scientists who work for them. Those industrial scientists who do report problems to the press are called "whistle-blowers," for they are explicitly defying corporate norms.[21] Corporate restrictions on communication are so embedded in the system, so much a part of the working assumptions of the press, that they are seldom challenged until an industrial disaster such as the Three Mile Island accident or the Bhopal chemical leak raises questions about the implications of corporate secrecy.

Recently federal and state governments have moved to increase restrictions on the flow of information to the press, for example, by imposing a system of prior censorship on government employees who wish to write about their work, and by reducing the number of press briefings. The government has also limited press access through changes in the Freedom of Information Act, budget constraints on the information-gathering activities of public agencies, and increased classification of federally funded research, even in areas that are not directly related to national security.[22]

Washington, according to reporters, still "leaks like a sieve." However, journalists who cover controversial issues often complain of the difficulty of obtaining information on sensitive topics. An environment reporter probing agencies in Washington and Albany for information on Hudson River pollution complained, "You can get all the press releases you want. The real problem is getting them to say something." Another reporter working on this story contacted the New York state public relations office. "They usually have the information I need. But I spent six hours on the phone, trying to get them to tell me how much they have spent on the [pollution control] project. They just wouldn't give me the information."[23]

Accidents at nuclear power plants have generated efforts by both industry and government to exclude the press. The sensational and inaccurate reporting of the Three Mile Island accident, which reflected as well as exacerbated the confusion that characterized this crisis, led to government and utility proposals to restrict press access to such events until official technical reports were available.[24] Several years later, in April 1986, the U.S. Department of Energy and the Nuclear Regulatory Commission responded to the Chernobyl accident in the Soviet Union by placing gag orders on their own employees and contractors, including scientists in national laboratories. They were to avoid the press or to limit their statements to background material. Despite their criticism of the Soviet Union for withholding details of the accident, these agencies also tried to limit public information. According to one memorandum, contractors and employees were not to make comparisons between U.S. and Russian reactors. Another voiced "a strong determination on the part of the U.S. government to speak with one voice" on matters relating to Chernobyl. All requests for information were to be channeled to three people.[25]

Restrictions on press access to events are sometimes imposed directly by scientists. During the controversy over recombinant DNA, research biologists tried to exclude reporters from the 1975 meeting at the Asilomar Conference Center on potential risks of this research. They eventually allowed sixteen reporters to attend, but set boundaries on the discussion. It was to focus on technical questions of risk, avoiding philosophical issues of creating life, political issues of research funding, and social issues of the implications of genetic engineering. An embargo was placed on all stories until the conference was over.[26]

Scientists have also proposed restricting press access to information about ongoing medical research.[27] They are concerned that press reports can have a harmful effect on individual behavior as well as medical practice. For example, in 1983 the *New England Journal of Medicine* published the preliminary results of research on a new therapy using hyperbaric oxygen for relief of the symptoms of multiple sclerosis. The scientific report was cautious, announcing that "preliminary results" suggested a "positive but transient effect," which warranted further study, and warning that the therapy could not generally be recommended. The press summaries of the report were cautious in tone and included these qualifications. Nevertheless, hospitals were still bombarded with requests from multiple sclerosis victims and their families. And special private clinics opened storefront operations to administer oxygen.[28]

Press reports about new scientific advances tend to raise the hopes of desperate people. Dr. Louis Lasagna talks about the "awful impact" of *Reader's Digest* on the practice of medicine as patients come to their doctors' offices brandishing the latest copy and demanding the latest cure.[29] Dr. Larry Lamb, a medical writer whose syndicated column appears in

500 daily newspapers, receives about 20,000 hopeful letters from readers every month, including some from physicians asking for information to convey to their patients.[30]

When scientific research bears directly on health, it is often difficult to judge when best to release information to the press. How much evidence is necessary? How certain must the evidence be? Consider, for example, one scientist's response to the 1979 press release issued by the National Cancer Institute (NCI) on a technical report showing that Dapsone, a drug used to treat Hansen's disease (leprosy), caused an increase of cancer in rats. The *New York Times* and other papers reported the findings. Barry R. Bloom, chairman of the Department of Microbiology at the Albert Einstein College of Medicine, criticized the NCI for releasing data from the study because the implications of animal tests for human beings were still questionable. Data, he argues, should not be published until carefully evaluated in terms of "the totality of available evidence." Bloom believes that "until data are interpreted and validated, until the experimental design and significance are reviewed, and until all currently available data on the incidence of cancer in exposed human populations can be integrated, the rush to the press is simply mindless if not unethical."[31] Taken literally, however, such constraints on releasing information to the public go beyond reasonable reticence; the press would wait forever for science news.

Scientists have also proposed control of public access to information by diverting reporters to professional intermediaries or contact officers, who will provide official versions. Richard Fiske, a geologist who directs the Museum of Natural History of the Smithsonian Institution, has proposed that working scientists be isolated from the media. Speaking at a professional society meeting, he described the chaotic reporting of the volcanic crisis at Guadalupe, an event marked by uncertainty, confusion, and scientific disagreement about the nature of risk. Reporters had talked directly to a number of scientists, each of whom offered a somewhat different interpretation of what was going on. Thus they discovered and reported the disagreements within the scientific community. Fiske's response was to suggest that in the future all press communications be made through a public relations officer, who would assure that only "depersonalized" and "consistent" information be released. His proposal met with enthusiastic applause from his colleagues at the meeting.[32] Similar proposals were discussed at a conference on medicine and the media. Participating medical researchers talked approvingly about establishing "truth squads" and "a central clearinghouse" that would verify scientific information before it was released to the press.[33]

Other scientists have proposed a related means of control that would create a new profession within the scientific community to mediate between science and the press. A physicist, concerned about public attitudes toward nuclear power and hoping to reduce "media distortion," submitted such a proposal to the Panel on Public Affairs of the American Physical Society. He wanted scientific societies to certify those scientists who were technically competent to speak to reporters.[34] Another scientist has proposed the creation of a profession of "certified public scientists" to make independent technical evaluations of scientific disputes and to be responsible for their public communication.[35]

These are paternalistic proposals; they are easily rationalized, often reasonable, but they place scientists in the inappropriate role of public guardian. As most efforts to censor information, they are also difficult to contain. And as the control of information at Chernobyl suggests, professional or political control over news can foster rumor, uninformed speculation, and sometimes unwarranted fear.

Official accounts, selected and structured to emphasize the positive aspects of an issue, may also cause costly delays in bringing a problem to public attention. For example, for years the press relied on official accounts about nuclear power, underreporting the concerns expressed by those outside the establishment. Simply acting as a conduit transmitting the views of their sources, journalists provided optimistic and positive views of a technology "too cheap to meter" that would solve the agricultural and energy problems of the world. Even just before the accident at Three Mile Island, the local press covered the development of this facility by using only the information provided by utility officials, who consistently emphasized "good news."[36]

Similarly, space reporters promoted the NASA programs, using the language and context provided by their technical informers. In each case, idealizations designed to preserve the status quo were passed on to the public as facts. The press acceptance of NASA's control of communications about the shuttle program led to one-sided publicity that ultimately undermined the credibility of both the press and the technical sources of information.

In the reporting of environmental problems there are many examples of efforts by scientists or public officials to control communication. These have had high social costs, delaying public recognition of serious problems and preventing appropriate responses. The reluctance of scientists to talk directly to journalists was, according to Frank Graham, Jr., responsible for the long delay in public awareness of the pesticide problem: "In the face of man's massive intervention in the functioning of the natural world, the scientific establishment simply filed the ominous facts and kept mum. . . . They sneered at such techniques as popularization, and recoiled in indignation from the suggestion that they cooperate with the mass media to put across the story that should have been told."[37]

The press in Michigan took over two years to report on food contamination when a fire retardant chemical, PBB, was accidentally mixed with cattle feed in 1973. Local reporters, who relied on state agencies as sources of expertise, were simply reassured that the problem was contained. The national press did not give significant attention to the problem until it became an issue in the 1976 election campaign.

The long delay in the press coverage of dioxin contamination at Times Beach can also be traced in part to efforts by public officials to minimize public knowledge. In 1974, faced with a damaging technical report, the director of Epidemiology and Disease Control at the Missouri Division of Health wrote a memorandum suggesting that "extreme care be exercised in any release of information for public consumption to the press."[38] The report was therefore not publicized. Not until eight years later did dioxin in Missouri finally become news, a delay with potentially serious implications for the health of local residents.

The efforts of scientists to manage the media—through both public relations or communication controls—reflect their continued ambivalence about the press. Scientists today see improved press coverage as a means of fulfilling their obligation to bring science to the public and attracting support from legislators, corporate leaders, and foundation executives. But they have also carried over values from a time when science was less accountable and more isolated from public affairs. Scientists fear that publicity will direct research funding toward applied and newsworthy areas at the expense of less dramatic but more productive fields. They worry about the corruptive influence on science of self-promotion and the encouragement of scientists more skilled in public relations than in research. They worry about distortion of information, inaccuracy, and sensationalism. Relations between science and the press remain fragile and strained. However, the question is no longer whether science will be covered in the press, but how it will be conveyed. While the two cultures remain in tension, they are also inextricably bound.

10

THE HIGH COST OF HYPE

On January 28, 1986, a long-standing and comfortable part-
nership between NASA and the press was shattered, when
the space shuttle Challenger exploded seconds after lift-off,
killing all aboard.[1] The press reaction to the explosion was
one of grief, disillusionment, and rage. For many longtime
space journalists the event was a personal tragedy. "Those
people were me," wrote a Houston reporter. "The shining
star of technology for 30 years has dimmed." The *Miami Herald* compared the "countdown to disaster" to a "Greek tragedy, peppered with portents of the doom to come." The *New
York Times* wrote of its disillusionment with an agency that

"has symbolized all that is best in American technology . . . computerized, at the cutting edge of technology, sophisticated in its public relations strategy, squeaky-clean in its integrity."[2]

The space program had been important to the development of science journalism as a profession. The many months at Cape Canaveral had brought together journalists interested in science and technology, and attracted new writers to the field. For 30 years they had covered the space program as an awesome and pioneering venture, a source of national prestige. The first space shuttle in 1981 assumed symbolic dimensions in the popular press as an affirmation of American faith in science and technology, a solution to problems of military security, a "sweet vindication of American know-how." In effect, the press reports of space launches incorporated all the images that are so characteristic of science and technology journalism.

Fascinated with the technology, reporters for years had simply accepted what NASA fed them, reproducing the agency's assertions, promoting the prepackaged information they received, and rarely questioning the premises of the program, the competence of the scientists, or the safety of the operation. Only three days before the accident, a *Boston Globe* reporter joked about NASA's public relations: "How does NASA spell publicity? Christa McAuliffe," referring to the school teacher who was among the astronauts. Three days later, McAuliffe was called "the victim of a PR campaign."

After the accident an angry press felt betrayed. *Newsweek* announced that "the news media and NASA, wedded by mutual interest from the earliest days of the space program, are in the midst of a messy divorce." Having suddenly lost faith in the veracity of NASA, some newspapers even engaged in electronic war games, using high-technology interception antennas and experimental laser cameras to get stories about the recovery of the shuttle that NASA wanted to conceal.[3] The press was filled with self-incrimination, as reporters accused themselves of accepting "spoon-fed news," of ignoring the safety problems of NASA by focusing only on the launches, of "treating the shuttle like a running photo opportunity," of letting readers down. More than any other event, the *Challenger* accident brought press and public awareness of the importance of probing and critical science journalism.

Science writers, in effect, are brokers, framing social reality for their readers and shaping the public consciousness about science-related events.[4] Through their selection of news about science and technology they set the agenda for public policy. Through their presentation of science news they lay the foundation for personal attitudes and public actions. For they are often our only source of information about the technical choices that significantly affect our lives.

Press coverage of science and technology is increasing, reflecting the pervasiveness of science and technology in business, politics, and health. Scientific and technological choices affect our work, our health, our lives. We pay for their implementation and bear their social costs. Public understanding of their social implications, their technical justifications, and their political and economic foundations is in the interest of an informed and involved citizenry. It is also critical to the health of our scientific and technological enterprise. The high cost of public naiveté regarding science and the nature of scientific evidence has been apparent in many controversies—over the value of animal experimentation, the appropriate precautions to prevent the spread of AIDS, the risks of a nuclear power plant explosion, and the teaching of evolution in the schools.

The press can play an important role in enhancing public understanding, but it frequently fails to do so. There are many examples of brilliant science reporting, written with analytic clarity, critical insight, and provocative style; but too often science in the press is more a subject for consumption than for public scrutiny, more a source of entertainment than of information.[5] Too often science is presented as an arcane activity outside and above the sphere of normal human understanding, and therefore beyond our control. Too often the coverage is promotional and uncritical, encouraging apathy, a sense of impotence, and the ubiquitous tendency to defer to expertise.

Science is practiced by an elite, but its impact extends to us all. Yet political questions of scientific responsibility and accountability are seldom considered news; nor are the ideologies or social priorities that guide science policy decisions. Focusing on individual accomplishments and dramatic or controversial events, journalists convey little about the sociology of science, the structure of scientific institutions, or the daily routines of research. We read about the results of research and the stories of success, but not about the process, the dead ends, the wrong turns. Who discovered what is more newsworthy than what was discovered or how. Thus science in the press becomes a form of sport, a "race" between scientists in different disciplines or between competitive nations.

There is little in this type of reporting to help the reader understand the nature of scientific evidence and the difference between science and unverified opinion. As a result, when new problems emerge as the focus of public concern, people are ill-prepared to deal with scientific information. The persistent fear of catching AIDS through casual contact with AIDS victims despite scientific evidence to the contrary is a case in point.

The reporting of technology, like that of science, tends to be promotional. Many writers convey a fervent conviction that new technology will create a better world. But the message is polarized—we read of either promising applications or perilous effects, of triumphant progress or tragic risks. Impending breakthroughs are reported with zeal, and technological failures are reported with alarm. But the long-term political and social consequences of technological choices are seldom explored. Thus technology in the press becomes a side show unrelated to events at center stage.

This study has suggested that many of these characteristics of science and technology reporting follow from the nature of the relationship between journalists and their sources. Many scientists today, concerned about their legitimacy in the political arena and anxious to receive support for their work, are sensitive to their image in the press. Hoping to shape that image, they are becoming adept at packaging information for journalists. Like advocates in any field, they are prone to overestimate the benefit of their work and minimize its risks. Indeed, the problems of science and technology reporting can often be traced to the influence of sources advocating their ideas.

For their part, journalists, especially those with limited experience in science reporting, are vulnerable to manipulation by their sources of information. They are concerned about balance and objectivity and accept the ideology of science as a neutral source of authority, an objective judge of truth. Some science writers are in awe of scientists; others are intimidated. But most are bewildered by the complexity of technical issues. The difficulty of evaluating a complex and uncertain subject converges with the day-to-day constraints of the journalistic profession to reinforce the tendency to rely uncritically on scientific expertise. While political writers often go well beyond press briefings to probe the stories behind the news, science reporters tend to rely on scientific authorities, press conferences, and professional journals. The result? Many journalists have adopted the mind-set or "frame" of scientists, interpreting science in terms defined by their sources, even when those sources are clearly interested in projecting a particular view.

Thus while art, theater, music, and literature are routinely subjected to criticism, science and technology are almost always spared. While political writers aim to analyze and criticize, science writers seek to elucidate and explain. Few are the outlets for journalists who would serve as critical commentators on, or probing investigators of, science and technology. Rare are the Walter Lippmanns or I. F. Stones of science who write regularly in the press.[6] Unaggressive in their reporting and relying on official sources, science journalists present a narrow range of coverage. Many journalists are, in effect, retailing science and technology more than investigating them, identifying with their sources more than challenging them.

If the reporting of science and technology is so uncritical, why is there continued tension between scientists and the press? The communities of science and journalism differ in certain fundamental and important respects. To begin with, they often differ in their judgments about what is news. In the scientific community, research results become reliable and therefore newsworthy through replication by and endorsement of professional colleagues. Prior to publication in reputable journals, scientific papers are carefully evaluated and approved through the system of peer review. This system of establishing reliability is critical to the structure of science, and especially to the process of scientific communication. For scientists, then, research findings are tentative, undigested, provisional—and therefore not newsworthy—until certified by peers to fit into the existing framework of knowledge. For journalists, on the other hand, certified and established ideas are "old news"—of far less interest than new and dramatic, though possibly tentative, research. Seeking to entertain as well as to inform, they are attracted to nonroutine, nonconventional, and even aberrant events.[7] This difference between scientists and journalists often becomes a source of contention when overzealous researchers seek press coverage of "hot" research prior to the time-consuming process of peer review.

In their search for credible perspectives on controversial

issues, journalists often rely on the opinions of scientists who have become well-known public figures. Nobel Prize winners are frequently cited in fields well outside their specialized expertise, journalists having sought their opinions simply because of their general prestige in science and the familiarity of their names. Scientists suspect such use of unverified opinion. Arnold Relman, editor of the *New England Journal of Medicine*, expressed the scientists' view: "If a [politician] makes a statement of what the policy of his government is or what he thinks or what he is going to vote, that's news. . . . News of a new development in science is coupled with evidence. Opinion is not important, it's evidence. Opinion is cheap and can be misleading in science, but opinion in politics or public affairs is another matter."[8]

Certain professional practices that are part of journalism conflict with scientific expectations about appropriate styles of communication. For example, while both groups are committed to communicating truth, journalists must often omit the careful documentation and precautionary qualifications that scientists feel are necessary to accurately present their work. While scientists are socialized to qualify their findings, journalists may see qualifications as protective coloration. Furthermore, readability in the eyes of the journalist may be oversimplification to the scientist. Indeed, many accusations of inaccuracy are traceable to reporters' efforts to present complex material in a readable and appealing style.

Journalistic conventions intended to enhance audience appeal may also violate scientific norms. For example, to make abstract technical decisions more concrete, science writers often examine the personal choices of their technical informants ("Would you live at Love Canal?"), undermining the idea that technical decisions are based on depersonalized evidence. To create a human interest angle, journalists also personalize science; but the focus on individual accomplishments and the presentation of scientists as stars contradicts communal norms, which favor a collective image of science as an objective and disinterested profession. Similarly, to convince their editors about the newsworthiness of science and technology, journalists tend to emphasize the uniqueness of individual events (the "first" discovery, the "breakthrough"). Although many scientists actively contribute to the breakthrough syndrome, ideally they prefer to emphasize continuity and the cumulative nature of research.

The journalistic preoccupation with conflict and aberration, intended to attract the reader's interest, is a further source of strain. In covering disputes journalists tend to create polarities: technologies are either risky or they are safe. The quest for simplicity, drama, and brevity precludes the complex, nuanced positions that scientists prefer. But the polarized presentation of technical disputes also reflects journalists' norms of objectivity—their belief that verity can be established by balancing conflicting claims. This approach further contributes to strain, for objectivity to a scientist is based on the understanding that claims must be verified by empirical means—hardly by balancing opposing views.

Differences in the use of language add to the strain. The language of science is intended to be precise and instrumental. Scientists communicate for a purpose—to indicate regularities and aggregate patterns, and to provide technical data. In contrast, journalistic language has literary roots. Journalists will choose words for their richness of reference, their suggestiveness, their graphic appeal. They are likely to prefer a "toxic dump" to a "waste disposal facility."

In any discourse, language is organized to address the background and assumptions of the anticipated audience.[9] Scientists direct their professional communication to an audience that is trained in their discipline. They take for granted that their readers share certain assumptions and therefore will assimilate the information conveyed in predictable ways.[10] Journalists, on the other hand, write for di-

verse readers who will interpret the information in subjective terms, depending on their interests, objectives, and
technical sophistication.[11] Thus, while scientists talk of aggregate data, reporters write of the immediate concerns of
their readers: "Should I use saccharin? Will I be harmed?"[12]

Often words that have a special meaning in a scientific
context will be interpreted differently by the lay reader. For
example, the word "epidemic" has both technical and general connotations. Scientists use the word "epidemic" to describe a cluster of incidents greater than the normal background level of cases. If the background level is zero, then
six cases are technically an epidemic. To the public or the
journalist an epidemic implies thousands of cases, a ram-
pantly spreading disease.

Confusion over the definition of "evidence" occurs
among scientists as well as in the press, often confounding
the discourse of risk disputes. Biostatisticians use the word
"evidence" as a statistical concept. But for biomedical researchers the critical experiment may also be defined as evidence. Most lay people accept as credible evidence anecdotal
information or individual cases. So, too, do journalists. Such
differences frequently lead to misunderstanding. In reports
about the health effects of exposure to toxic chemicals at
Love Canal and Times Beach, for example, scientists and
journalists held different assumptions—about the definition
of credible evidence concerning the validity of animal tests,
the neighborhood's habitability, and the adequacy of containment of the chemicals. Thus, when scientists described
the health effects of dioxin with a cryptic "no evidence,"
meaning no statistically significant evidence, journalists interpreted their response as an effort to cover up the problem,
since they knew of individual cases.

Similar linguistic confusion marked the dispute over the
report written by the National Academy of Sciences panel
on food additives, when the Academy's placement of saccha-

rin in a "moderate risk category" was interpreted by the press to mean it was a "moderate cancer-causing agent."

Perhaps the most important source of strain between scientists and journalists lies in their differing views about the appropriate role of the press. Scientists often talk about the press as a conduit or pipeline, responsible simply for transmitting science to the public in a way that it can be easily understood. They expect to control this flow of information to the public as they do within their own domain. Confusing their special interests with general questions about the responsibility of the press, they are reluctant to tolerate independent analysis of the limits or flaws of science. They assume that the purpose of journalism is to convey a positive image that will promote science, and they see the press as means of furthering scientific goals.

This view of journalism is reflected in scientists' complaints about the press and its effects on public attitudes. Scientists tend to attribute negative public attitudes about science and technology to problems of media communication, ultimately to journalists, who, they believe, distort the flow of information from scientists to the public. Alternatively, however, problems of scientific communication could as easily be attributed to the sources of information, to suppression of facts, to manipulation of information, or to overeager, promotional public relations.[13]

Many science journalists, of course, have a perception of their role that is not too different from that of scientists. They see their mission as one of recording "official history"—of elucidating and even eulogizing science. But there are some who are beginning to question their role as "self-appointed trumpets" for science and technology. Reacting to events such as Three Mile Island, Love Canal, or the Challenger explosion, and to the economic implications of large and costly scientific endeavors, they are beginning to suspect promotional hype about science and technology, and to raise probing questions in their interviews with scientists: Who pays? Who is responsible? What's in it for the public? What are the stakes?

While "gee whiz," "cosmic breakthrough" articles continue to dominate press coverage of science and technology, a number of journalists today want to probe scientific issues so that, as one journalist put it, "public expectations do not get out of control." "It is not enough for us to report the new discoveries or gadgetries; we must delve deeper into their effects on people and public policy." "I want to take some of the awesomeness out of science." "I want to create a better-informed citizenry able to deal with problems." These are among the goals expressed by at least some journalists today.

Efforts to improve the standards of science journalism are also apparent in the increased professionalization of the field. The professional organization of science writers is the National Association of Science Writers (NASW), founded in 1934 by a dozen veteran science writers[14] to "foster the dissemination of accurate scientific knowledge by the press of the nation in cooperation with scientific organizations and individual scientists."[15] Convinced of the public importance of science, these writers were concerned about the minimal level of press coverage. Struggling to convince their editors that science was news, the NASW founders hoped that professionalizing their specialty would enhance its visibility, recognition, and prestige. Moreover, a professional society could provide credentials to improve reporters' relations with scientists. Said David Dietz, one of the founders and Scripps Howard's science editor, "Membership in the NASW would be the kind of credential a scientist would appreciate."[16]

The NASW remained a very small organization until after World War II, when press coverage of science and technology began to expand. Reflecting this expansion, it grew from 113 members in 1950 to 413 members in 1960 and 830 in 1970. It has 1200 members today. The organization brings together prospective employers and science writers, encourages recruitment into the specialty, and facilitates contacts with reliable scientific sources of information. Its newsletter disseminates professional "gossip," information on NASW activities and members, and special articles of interest to the advancement of the profession.[17]

In 1960 NASW spawned an independent nonprofit organization, the Council for the Advancement of Science Writing (CASW), a group whose membership includes writers, editors, educators, and scientists. The council raises funds to support the development of curriculum in schools of journalism, as well as annual briefings (the "New Horizons in Science" program) in which distinguished scientists talk about current scientific advances. The CASW also supports regional science reporting workshops, a fellowship and internship program, and special courses for journalists on such topics as biostatistics.

Formal training in science is increasingly viewed as essential background for science journalists, and special courses have proliferated. There are about 43 programs in science journalism in 67 colleges and universities. Fourteen offer masters degrees in the field. These programs include science requirements, so that soon most younger reporters specializing in science writing will have some science background.

Other efforts have been initiated by scientists. In order to reduce mutual suspicion between scientists and journalists and to assure that timely and reliable information reaches the press, the Scientists' Institute for Public Information (SIPI) has organized a Media Resource Service (MRS). As a clearinghouse, MRS includes a computer file of about 20,000 scientists and engineers who have agreed to answer queries from reporters. It receives over 50 telephone calls a week from journalists who are seeking reliable sources on a very wide range of subjects that call for scientific information. When a crisis such as the Chernobyl accident or the space shuttle explosion occurs, hundreds of reporters will call. If the subject of inquiry is a disputed one, MRS routes journalists to several scientists selected to represent a spectrum of opinion. The organization also brings together scientists and journalists in round table discussions of controversial scientific issues such as animal experimentation or the disposal of toxic wastes. Their purpose is to enhance both the technical sophistication of journalists and the political sophistication of scientists involved in these disputes.[18]

The success of MRS suggests that a significant number of scientists are willing to work for better relations. But tension between science and journalism is likely to persist. For the differences between these two communities are fundamental—following, as they do, from the institutional constraints and external pressures that each profession must face. Improving the scientific knowledge of journalists and enhancing the political understanding of scientists are both important, but maintaining their differences is also essential if each community is to fulfill its unique social role.

Indeed, the tension can itself be healthy. If the popular press is to play its traditional role as a watchdog over major social and political institutions, if it is to mediate between science and the public and facilitate the public discourse about crucial policy issues, both scientists and journalists must accept and come to terms with an uneasy and often adversarial relationship. Scientists must restrain the promotional tendencies that lead to controls on information or to oversell, and they must open their doors to more probing investigation. And journalists on their part must try to convey understanding as well as information. It is not enough to merely react to scientific events, translating and elucidating them for popular consumption. To understand science and technology, readers need to know their context: the social, political, and economic implications of scientific activities, the nature of evidence underlying decisions, and the limits as well as the power of science as applied to human affairs.

NOTES

PREFACE

1. Television is obviously an increasingly important source of information. So, too, are science museums. But the print media continues to be the primary source of news about science and technology. Studies of the relative importance of various media include Serena Wade and Wilbur Schramm, "The Mass Media as Sources of Public Affairs, Science and Health Knowledge," *Public Opinion Quarterly* 33, summer 1969, pp. 197–209; Lawrence W. Lichty, "The News Media," *Wilson Quarterly* 6, 1982, pp. 49–57; and Evan Witt, "Here, There and Everywhere: Where Americans Get Their News," *Public Opinion* 6, August/September 1983, pp. 45–48.

2. This program is run by David Rubin for the School of Journalism at New York University.

Chapter 1. IMAGES OF SCIENCE AND TECHNOLOGY

1. William Laurence, oral history archives, Columbia University, quoted in Spencer Weart, *Nuclear Fear: A History of Images* (New York: Doubleday, in press), chap. 5.

2. Harold Schmeck and Walter Sullivan were the other two science journalists. Both are still writing for the *New York Times* in 1986.

3. Frank Carey, "A Quarter Century of Science Reporting," *Neiman Reports* 20, June 1966, pp. 7–8.

4. For a review of the history of the research developments, see Sandra Panem, *The Interferon Crusade* (Washington, D.C.: Brookings, 1984).

5. The tone of coverage in magazines and newspapers can be traced through indices in the *Readers Guide* and *Newsbank*.

6. Harold Schmeck's controversial article appeared in the *New York Times* on May 27, 1980. The reply appeared on June 17, 1980.

7. See discussion of scientists' criticism of the press in June Goodfield, *Reflections on Science and the Media* (Washington, D.C.: AAAS, 1981). See also Jay A. Winsten, "Science and the Media: The Boundaries of Truth," *Health Affairs* 4, spring 1985, pp. 5–23. Some excellent articles on the relationship between scientists and journalists appear in Sharon Friedman, Sharon Dunwoody, and Carol Rogers (eds.), *Scientists and Journalists* (New York: Free Press, 1986).

8. William Burrows, "Science Meets the Press: Bad Chemistry," *Sciences*, April 1980, pp. 15–19.

9. In a study of Washington reporters, Stephen Hess asked them which papers they read regularly. He found that 89 percent read the *Washington Post*, 73 percent the *New York Times*, and 51 percent the *Wall Street Journal*. Hess also surveyed public officials. Of his sample of high federal officials 90 percent read the *Post*, 45 percent the *Times*, and 62 percent the *Wall Street Journal*. Stephen Hess, *The Washington Reporters* (Washington, D.C.: Brookings, 1981).

10. The press today also includes the alternative or political newspapers which assume an explicit advocacy role and often focus on specific policy issues. There are also a growing number of suburban and community newspapers with little writing about science. I am not including these newspapers in this study. A lot of science news appears in special sections of scientific journals such as *Science* and the *New England Journal of Medicine*. These are directed to the scientific community, not the lay public, and are not included in this study. Similarly, I am excluding specialized science magazines such as *Science*, *Science 86* and *Scientific American* which are more devoted to popularizing science than to reporting science news, or which direct their writing to a specialized audience interested in, and often well informed about, scientific issues.

11. Todd Gitlin, *The Whole World Is Watching* (Berkeley: University of California Press, 1980), p. 7. See also Gaye Tuchman, *Making News* (New York: Free Press, 1978).

12. See William Gamson and Kathryn Lasch, *Evaluating the Welfare State* (New York: Academic Press, 1983).

13. Edward Lawless, *Technology and Social Shock* (New Brunswick: Rutgers University Press, 1977); Herbert Gans, *Deciding What's News* (New York: Vintage Books, 1980); Michael Schudson, *Discovering the News* (New York: Basic Books, 1978); and Warren Breed, *The Newspapermen: News and Society* (New York: Arno Press, 1980).

14. Kenneth Burke, "Literature as Equipment for Living," *The Philosophy of Literary Form* (Berkeley: University of California Press, 1973), p. 298.

15. George Lakoff and Mark Johnson, *Metaphors We Live By* (Chicago: University of Chicago Press, 1980), suggest how metaphors define reality. See also Murray Edelman, *Political Language* (New York: Academic Press, 1977).

Chapter 2. THE MYSTIQUE OF SCIENCE IN THE PRESS

1. *Nation,* January 16, 1902.

2. *Kansas City Times,* September 17, 1982.

3. *Time,* October 27, 1980.

4. *Newsweek,* October 27, 1979.

5. *Time,* October 22, 1979.

6. *McCall's,* July 1964, p. 40.

7. *New York Times,* February 19, 1983.

8. *New York Times,* June 12, 1984.

9. *Binghampton Sunday Press,* March 14, 1982.

10. *McCall's,* July 1964, pp. 38–40, 124.

11. *Science Digest* 55, February 1964, pp. 30–36.

12. *Time,* November 6, 1964.

13. *Vogue,* January 1978, p. 174.

14. *Family Health,* June 1978, p. 24.

15. *Newsweek,* October 24, 1983.

16. *New York Times,* October 11, 1983.

17. *Time,* October 29, 1979.

18. Ibid.

19. *U.S. News and World Report,* April 16, 1981.

20. See, for example, *Time,* December 6, 1982, and the *Christian Science Monitor,* education section, April 15, 1983; also see the local papers catalogued in *Newsbank.*

21. *Milwaukee Journal,* July 11, 1982.

22. *San Jose Mercury,* March 12, 1982.

23. The quotation by Glenn Seaborg appeared in the *Houston Post,* November 17, 1982, and many other newspapers.

24. *Business Week,* March 28, 1983; *Seattle Times,* January 2, 1983.

25. *Business Week* and many newspapers emphasize the problem of obsolete equipment and reinforce this with photographs.

26. Op ed piece by William McCowan, "Illiterassee att Wurk," *New York Times,* August 19, 1982, on the "unlettered underclass."

27. *Time,* December 6, 1982.

28. *Boston Globe,* June 19, 1980.

29. *New York Times,* August 5, 1980.

30. *Christian Science Monitor,* March 10, 1982.

31. *Newsweek,* February 8, 1982.

32. Edward O. Wilson, *On Human Nature* (Cambridge, Mass.: Harvard University Press, 1980).

33. For a comprehensive review of the criticism and the controversy, see Ullica Segerstrale, "Colleagues in Conflict: An In Vivo Analysis of the Sociobiology Controversy," *Biology and Philosophy* 1, 1986, pp. 53–87.

34. *Playboy,* August 1978.

35. *Time,* August 1, 1977.

36. *Cosmopolitan,* March 1982.

37. Camille Benbow and Julian Stanley, "Sex Differences in Mathematical Reasoning: Fact or Artifact?" *Science* 210, December 12, 1980, pp. 1262–1264.

38. *New York Times,* December 7, 1980.

39. *Time,* December 15, 1980.

40. Pamela Weintraub, "The Brain: His and Hers," *Discover,* April 1981, pp. 15–20.

41. See, for example, the *New York Times,* May 30, 1981.

42. *Business Week,* April 10, 1978.

43. See, for example, interviews in the *New York Times,* October 12, 1975, and *People Weekly,* November 19, 1975.

44. *Science News,* November 19, 1975.

45. *Newsweek,* April 12, 1976.

46. *Science Digest,* March 1982.

Chapter 3. THE PRESS ON THE TECHNOLOGICAL FRONTIER

1. *U.S. News and World Report,* December 28, 1981; January 4, 1982.

2. *Salt Lake Tribune,* March 1983.

3. *U.S. News and World Report,* September 15, 1980.

4. *Christian Science Monitor,* June 9, 1982.

5. *Columbus Evening Dispatch,* January 23, 1983.

6. *Newsweek,* October 18, 1957.

7. *Newsweek,* August 9, 1982.

8. *Kansas City Star,* January 9, 1983.

9. *Salt Lake Tribune,* March 14, 1983.

10. *Birmingham Alabama News,* October 27, 1982.

11. Such advertisements appear, for example, in *New York Times* high-technology career supplements.

12. *Newsweek,* March 7, 1983, p. 67.

13. *New York Times,* September 18, 1983.

14. *Seattle Times,* January 9, 1983.

15. Kenneth Wilson, cited in *SIPIScope* 13, November/December 1985, p. 11.

16. For coverage of the conference, see Sheldon Krimsky, *Genetic Alchemy* (Cambridge, Mass.: MIT Press, 1982) and Michael Rogers, *Biohazard* (New York: Knopf, 1977). For a review of news coverage, see Michael Altimore, "The Social Construction of a Scientific Controversy," *Science, Technology and Human Values,* fall 1982, pp. 24–31.

17. See, for example, the use of the language of miracles in a featured article, "The Miracle of Spliced Genes," *Newsweek*, March 17, 1980, and in *U.S. News and World Report,* December 12, 1979.

18. Liebe Calaviere, "New Strains of Life or Death?" *New York Times Magazine,* August 22, 1976.

19. *New York Times,* May 31, 1981.

20. From titles in the *Reader's Guide.*

21. See, for example, *Birmingham Alabama News,* May 25, 1983.

22. These quotes are from the local newspapers that are catalogued by issue in *Newsbank* 1982 and 1983.

23. *Life,* July 14, 1958.

24. *Life,* February 17, 1958.

25. See, for example, *Life,* September 17, 1971.

26. Lawrence K. Altman, "After Barney Clark: Reflections of a Reporter on Unsolved Issues," in Margery Shaw (ed.), *After Barney Clark* (Austin: University of Texas Press, 1984), pp. 113–128.

27. *New York Times,* December 3, 1982; and *Philadelphia Inquirer,* December 5, 1982.

28. *Salt Lake Tribune,* December 2, 1982.

29. *Salt Lake Tribune,* December 3, 1982.

30. *Salt Lake Tribune,* December 3, 1982.

31. *Salt Lake Tribune,* May 25, 1982.

32. In Harry Schwartz, "Towards the Conquest of Heart Disease," *New York Times Magazine,* March 27, 1983.

33. While writing this section, I discovered an excellent unpublished list of media reports and articles on ERT, compiled by Nancy T. Sommers, Princeton, New Jersey.

34. For example, see *San Diego Union*, December 13, 1964; *Pittsburgh Post Gazette*, November 10, 1964; and *Ladies' Home Journal*, January 1965.

35. Robert A. Wilson, "The Roles of Estrogen and Progesterone in Breast and Genital Cancer," *Journal of the American Medical Association*, 182, October 1962, pp. 327–331; and *Feminine Forever*, New York: M. Evans, 1966.

36. AP report, April 14, 1965, quoting Francis P. Rhoads.

37. AP report, January 29, 1966.

38. *Look*, January 11, 1966.

39. Robert Kistner developed his ideas in a popular book entitled *The Pill: Fact and Fallacies* (New York: Delacorte Press, 1969) and in the *Ladies' Home Journal* and other women's magazines in 1969.

40. *McCall's*, October 1971.

41. *Vogue*, January 1974.

42. *New York Times*, December 5, 1975.

43. Letter from Stanley Sauerhaft, Hill and Knowlton, to William Davis, Ayerst Laboratories, December 17, 1976.

44. Elliot Valenstein, *Great and Desperate Cures* (New York: Basic Books, 1986) describes the press coverage of psychosurgery in the 1940s.

45. Waldemar Kaempffert, "Science in the News: Psychosurgery," *New York Times*, January 11, 1942.

46. See, for example, articles in *Time*, July 24, 1978; July 31, 1978; August 7, 1978; February 19, 1979; and in *Newsweek*, July 24, 1978.

47. See discussion of these debates and their press coverage in Dorothy Nelkin and Chris Anne Raymond, "Tempest in a Test Tube," *The Sciences* 20, November 1980.

Chapter 4. THE PERILS OF PROGRESS

1. For a detailed analysis of the coverage of risk in the press, using examples from the nuclear debate, Three Mile Island, the swine flu vaccine, and Love Canal, see Dorothy Nelkin, Background paper for 20th Century Fund, *Science in the Streets* (New York: Priority Press, 1984).

2. J. E. Lovelock, R. J. Maggs, and R. J. Wade, "Halogenated Hydrocarbons in and over the Atlantic," *Nature* 241, January 19, 1973, pp. 194–196.

3. Lydia Dotto and Harold Schiff, *The Ozone War* (New York: Doubleday, 1978).

4. Mario J. Molina and F. Sherwood Rowland, "Stratospheric Sink for Chlorofluoromethanes: Chlorine Atom-Catalysed Destruction of Ozone," *Nature* 249, June 28, 1974, pp. 810–812.

5. These and similar articles appeared throughout 1975. See Dotto and Schiff, op. cit., chap. 7.

6. Cited in Dotto and Schiff, op. cit., p. 251.

7. Ibid., p. 169.

8. See discussion in Paul Brodeur, "Annals of Chemistry," *New Yorker*, June 9, 1986, pp. 70–87.

9. *New York Times*, September 14, 1976; *Washington Post*, September 14, 1976.

10. *Business Week*, October 18, 1969.

11. *Newsweek*, March 12, 1979.

12. The *Baltimore Sun*, April 9, 1977, reprinted this debate.

13. *Time*, August 26, 1974.

14. Daniel Greenberg, "Washington Report," *New England Journal of Medicine*, 300, March 22, 1979, pp. 687–688.

15. *New York Times*, February 28, 1979.

16. *New York Times*, March 3, 1979.

17. *Washington Post*, March 3, 1979. See also March 5 and 7, 1979.

18. See, for example, reports in *Newsweek* and *Time* during March and April 1977.

19. *Newsweek*, April 4, 1977. See also March 21, 1977.

20. *Time*, April 25, 1977.

21. See Adeline Levine, *Love Canal: Science, Politics and Rights* (Lexington, Mass.: D. C. Heath, 1982).

22. *Time*, September 22, 1980.

23. *Los Angeles Times*, May 9, 1983.

24. *Milwaukee Journal*, June 19, 1983.

25. *Atlanta Journal*, June 4, 1983.

26. *Arkansas Gazette*, June 17, 1983.

27. *Time*, May 2, 1983.

28. A Stanford University study of perceptions of bias in the media found that partisans systematically view media coverage of controversial events as biased against them. Indeed, partisans see different content in the same material. See Robert Vallone, Lee Ross, and Mark Lepper, "Biased Perception and Perceptions of Media Bias," unpublished paper, Stanford University, September 22, 1983.

Chapter 5. MEDIA MESSAGES, MEDIA EFFECTS

1. Reported by Eliot Marshall, "Public Attitudes to Technological Progress," *Science* 205, July 20, 1979, pp. 281–285.

2. See discussion and interviews by John Noble Wilford in the *New York Times*, May 5, 1986.

3. For a review of the efforts to establish the impact of the media see Denis McQuail, "The Influence and Effects of Mass Media" in

James Curran, M. Gurevich, and J. Woollacott (eds.), *Mass Communication and Society* (Beverly Hills, Calif.: Sage, 1979), chap. 3.

4. Charles Cooley, *Human Nature and the Social Order* (New York: Scribner, 1922). The work on communication by Cooley and his colleagues is reviewed in Daniel Czitrom, *Media and the American Mind* (Chapel Hill: University of North Carolina Press, 1982), chap. 4.

5. Walter Lippmann, quoted in the *Conservationist* 33, May/June 1979, p. 1.

6. Paul Lazarsfeld, Bernard Berelson, and Hazel Gaudet, *The People's Choice* (New York: Duell, Sloan and Pierce, 1944).

7. Raymond and Alice Bauer, "America, Mass Society and Mass Media," *Journal of Social Issues* 16, 1960, pp. 3–66.

8. Joseph Klapper, *The Effects of Mass Communication* (Glencoe, N.Y.: Free Press, 1960).

9. Marshall McLuhan, *The Gutenberg Galaxy* (Toronto: University of Toronto Press, 1962). See also discussion in Czitrom, op. cit., chap. 6.

10. George Gerbner *et al.*, "Scientists on the TV Screen," *Society*, May/June 1981, pp. 41–44. See also Marcel LaFollette, "Science on Television: Influences and Strategies," *Daedalus*, fall 1982, pp. 183–197; Jonathan Weiner, "Prime Time Science," *Sciences*, September 1980, pp. 6–13; and Nancy Signorielli, "Science and Television: Some Images and Viewer Conceptions of Social Reality," paper presented at the annual meeting of the American Academy of Arts and Sciences, Detroit, May 1983 (mimeo).

11. George Basalla, "Pop Science: The Depiction of Science in Popular Culture," in Gerald Holton and William Blanpied (eds.), *Science and Its Public* (Dordrecht: D. Reidel, 1976), p. 266.

12. Marcel LaFollette, *Authority, Promise and Expectation: The Image of Science and Scientists in American Popular Magazines, 1910–1955*, Ph.D. thesis, University of Indiana, 1979, pp. v, xiv.

13. The columns advocating estrogen replacement therapy are an example, as are books and columns on cancer cures, diet fads, and scientific advice on sexual techniques.

14. For a review of this coverage, see Curtis MacDougall, *Superstition and the Press* (Buffalo, N.Y.: Prometheus Books, 1983).

15. See Donald N. Leff, "Four Wondrous Weeks of Science and Medicine in the Amazing, Incredible Supermarket Press," *NASW Newsletter*, January 1980.

16. National Science Board, *Science Indicators 1980* (Washington, D.C.: Government Printing Office, 1980), pp. 158–181. The studies defined the "attentive public" as those individuals who display a high level of interest in science, an adequate level of current information, and a commitment to information acquisition that would assure continuing knowledgeability. The data from this report have been updated to 1983 with information through personal communication with Jon Miller, a principal investigator of the research.

17. National Cancer Institute, *Cancer Prevention Awareness Survey*, NIH #84-26-77 (Washington, D.C.: Government Printing Office, 1984), pp. 58, 64.

18. See discussion in Chapter 4.

19. Colin Seymour-Ure, *The Press, Politics and the Public* (London: Methuen, 1968); and James B. Lemert, *Does Communication Change Public Opinion After All?* (Chicago: Nelson-Hall, 1981), pp. 44–45.

20. Oscar Handlin, "Science and Technology in Popular Culture," *Daedalus* 94, winter 1965, p. 156.

21. Gilbert Omenn, "Are Kids Afraid to Become Scientists?" *Science* 83, September 1983, p. 18.

22. Jon Miller, K. Prescott, and R. Pearson, *The Attitudes of the U.S. Public Towards Science and Technology* (Chicago: National Opinion Research Center, 1980). See also Daniel Yankelovitch, "Changing Public Attitudes to Science and the Quality of Life," *Science, Technology and Human Values* 39, spring 1982, pp. 23–29.

23. Louis Harris poll, 1980.

24. Lydia Dotto and Harold Schiff, *The Ozone War* (New York: Doubleday, 1978).

25. *New York Times,* December 9, 1984.

26. Gallup poll of June 1985, reported in *New York Times,* October 19, 1985.

27. Elise F. Jones, J. R. Beniger, and C. Westoff, "Pill and IUD Discontinuation in the United States, 1970–1975: Influence of the Media," *Family Planning Perspectives* 12, November/December 1980, pp. 293–300.

28. Edward J. Robinson, "Analyzing the Impact of Science Reporting," *Journalism Quarterly* 40, 1963, pp. 306–314.

29. Ronald Troyer and Gerald Markle, *Cigarettes: The Battle over Smoking* (New Brunswick, N.J.: Rutgers University Press, 1983).

30. This "agenda setting" role has been widely documented. See M. E. McCombs and D. L. Shaw, "The Agenda Setting Function of the Mass Media," *Public Opinion Quarterly* 36, summer 1972, pp. 176–187; Bernard C. Cohen, *The Press and Foreign Policy* (Princeton: Princeton University Press, 1963); Graham Murdock, "Mass Communication and the Construction of Meaning," in Nigel Armstead (ed.), *Reconstructing Social Psychology* (Hammondsworth, England: Penguin, 1974); and Gaye Tuchman, *Making News* (New York: Free Press, 1978).

31. Allan Mazur and K. M. Waba, "The Mass Media at Love Canal and Three Mile Island," paper presented at the annual meeting of the American Sociological Association, Detroit, September 2, 1983. Mazur contends that the quantity of coverage of a controversial technology, regardless of the bias, creates negative attitudes. See Allan Mazur, *The Dynamics of Technical Controversy* (Washington, D.C.: Communications Press, 1981).

32. Gladys Lang and Kurt Lang, *The Battle for Public Opinion* (New York: Columbia University Press, 1983).

33. Harold Laski, *The American Democracy* (New York: Viking Press, 1984).

34. Adeline Levine, *Love Canal: Science, Politics and People* (Lexington, Mass.: D. C. Heath, 1982).

35. Sheldon Krimsky, *Genetic Alchemy* (Cambridge, Mass.: MIT Press, 1983).

36. Gerald Markle (ed.), *The Politics of Laetrile* (Beverly Hills, Calif.: Sage, 1982).

37. Joseph D. Cooper, "Cyclamate Sequel: Risks Dominating Benefits," *Medical Tribune,* February 26, 1970, p. 11.

38. Sandra Panem, *The Interferon Crusade* (Washington, D.C.: Brookings, 1984).

39. The case, and the role of the press, is documented in Steven Maynard-Moody, "The Fetal Research Dispute," in Dorothy Nelkin (ed.), *Controversy: The Politics of Technical Decisions,* 2nd ed. (Beverly Hills, Calif.: Sage, 1984), pp. 213–232.

40. Dorothy Nelkin and Judith P. Swazey, "Science and Social Control," in Ruth Macklin (ed.), *Research on Violence* (New York: Plenum, 1981).

41. The influence of the public on the direction of research is a theme in Terry Shinn and Richard Whitley (eds.), *Expository Science* (Dordrecht: D. Reidel, 1985).

Chapter 6. THE CULTURE OF SCIENCE JOURNALISM

1. Quoted in Carolyn Hay, *A History of Science Writing in the United States,* master's thesis, Northwestern University, 1970, p. 35.

2. There are very few historical accounts of science journalism. See Hay, ibid.; David J. Rhees, *A New Voice for Science: Science Service under Edwin E. Slosson, 1921–29,* master's thesis, University of North Carolina, 1979; also Marcel LaFollette, *Authority, Promise and Expectation: The Images of Science and Scientists in American Popular Magazines, 1910–1955,* Ph.D. thesis, Indiana University, 1979. Frank Luther Matt in his classic history of magazines (*A History of American Magazines,* vols. 1–5, Cambridge, Mass.: Harvard Press, Belknap

Press, 1968; originally published, 1930) has useful reviews of science journalism.

3. On the popularization of physics, see Daniel Kevles, *The Physicists* (New York: Knopf, 1978), chap. 2.

4. Hay, op. cit., p. 34.

5. Quoted in LaFollette, op. cit., p. 189.

6. Frederick Lewis Allen, *Only Yesterday: An Informal History of the Twenties* (1931, reprint ed. New York: Harper & Row, 1964), pp. 164–165.

7. Quoted in Ronald Tobey, *The American Ideology of National Science: 1919–1930* (Pittsburgh: University of Pittsburgh Press, 1971), p. 106.

8. Hay, op. cit.

9. Quoted in Kevles, op. cit., p. 171.

10. Rhees, op. cit., pp. 31–32, and Tobey, op. cit.

11. Science Service, "A Statement of Purpose," *Science News Letter*, August 11, 1928, p. 90.

12. From Slosson's "Talks to Trustees," quoted in Rhees, op. cit., pp. 38–39.

13. Edwin Slosson, *Chats on Science* (New York: Century, 1924), p. 169.

14. Edwin Slosson, "Democracy of Knowledge," in B. Brownell (ed.), *Preface to the Universe* (New York: Van Nostrand, 1929), p. 108.

15. William Laurence, cited in Spencer Weart, *Nuclear Fear* (New York: Doubleday, in press), chap. 5.

16. Waldemar Kaempffert, *New York Times*, September 22, 1919; cited in Weart, op. cit., chap. 1.

17. Quoted in obituary for Gobind Lal, *NASW Newsletter*, April 1982.

18. Remarking on the concern about objectivity, Herbert J. Gans,

Deciding What's News (New York: Random House, 1979), p. 184, calls journalism "the strongest remaining bastion of logical positivism in America."

19. William L. Rivers, Wilbur Schramm, and Clifford Christian, *Responsibility in Mass Communication* (New York: Harper & Row, 1980), Appendix A.

20. Ibid., Appendix B.

21. Dan Schiller, *Objectivity and the News* (Philadelphia: University of Pennsylvania Press, 1981).

22. Ibid., pp. 80, 87.

23. Quoted in ibid., p. 180.

24. Quoted in Michael Schudson, *Discovering the News* (New York: Basic Books, 1978), pp. 78–80.

25. Quoted in Schiller, op. cit., p. 194.

26. See discussion in Edward Jay Epstein, *News from Nowhere* (New York: Random House, 1973).

27. Hay, op. cit.

28. Gaye Tuchman, "Objectivity as Strategic Ritual," *American Journal of Sociology* 77, 1972, pp. 660–679, and Tuchman, *Making News* (New York: Free Press, 1978).

29. Todd Gitlin, *The Whole World Is Watching* (Berkeley: University of California Press, 1980).

30. Harvey Molotch and Marilyn Lester, "News as Purposive Behavior," *American Sociological Review* 39, February 1974, pp. 101–112.

31. Interview with *Science* reporter Luther Carter.

32. *NASW Newsletter*, March 1966.

33. John Lear, "The Trouble with Science Writing," *Columbia Journalism Review* 9, 1972, p. 30.

34. David Perlman, "Science and the Mass Media," *Daedalus* 103, summer 1974, p. 207.

35. This was also the time of considerable expansion of the advocacy or political press, which denied all pretense of objectivity. See Chris Anne Raymond, *Uncovering Ideology: Occupational Health in the Mainstream and Advocacy Press*, Ph.D. thesis, Cornell University, 1983.

36. Perlman sent me his files of all his bylined articles for three years, representing three periods in his career. Obviously, because journalists write about what is going on, no selection of years can be used as typical of a writer's interests. For example, in 1976–1977 Perlman and other science reporters were preoccupied with genetic engineering. Other years nuclear power issues dominated their columns. I selected three years to suggest the range and tone of reporting at different times.

37. Personal correspondence with Perlman, January 24, 1984.

38. Quoted in Scott DeGarmo, "An Editor Takes a Survey: Are Scientists Better Science Writers Than Non-Scientists?" *NASW Newsletter*, October 1981, pp. 16–18; see also Pierre C. Fraley, "The Education and Training of Science Writers," *Journalism Quarterly* 40, 1963, pp. 323–328.

39. Nicholas Wade, "Scientists and the Press: Cancer Scare Story That Wasn't," *Science* 174, November 1971, pp. 679–680.

40. For background on the education of science writers, see Hillier Krieghbaum, *Science and the Mass Media* (New York: New York University Press, 1967), chap. 6, and Lee Z. Johnson, "Status and Attitudes of Science Writers," *Journalism Quarterly* 34, spring 1957, pp. 247–251.

41. From testimony in 1974 to the House of Representatives Select Committee on Labor; cited in an editorial in *Columbia Journalism Review*, September/October 1977, pp. 8–9.

42. Bob Hall, "The Brown Lung Controversy," *Columbia Journalism Review*, March/April 1978, p. 33.

43. Wade Roberts, "Phosvel: A Tale of Missed Cues," *Columbia Journalism Review*, July/August 1977, p. 23.

44. Betty Medsger, "Asbestos: The California Story," *Columbia Journalism Review,* September/October 1977, pp. 41–47.

45. Quoted in Hall, op. cit., p. 35.

46. Gans, op. cit. See also W. C. Johnstone et al., *The News People* (Urbana: University of Illinois Press, 1976).

47. *NASW Newsletter,* August 1979, p. 4.

48. Rae Goodell, "How to Kill a Controversy: The Case of Recombinant DNA," in Sharon M. Freidman, S. Dunwoody, and Carol Rogers (eds.), *Scientists and Journalists* (New York: Free Press, 1986), pp. 170–181.

49. David Perlman, self-portrait, in *Science in the Newspaper* (Washington, D.C.: AAAS, 1974), p. 8.

50. Ibid.

51. Victor McElheny, "Reporting of a Golden Age of Discovery," paper presented at the Conference on Expository Science, Paris, December 1–3, 1983, pp. 2–3 (mimeo).

52. Ed Edelson, "Science Writing Celebrates the Trivial," *NASW Newsletter,* December 1980.

53. These quotations and others without citation are from interviews, informal conversations with science writers, and their comments at meetings and press conferences.

Chapter 7. CONSTRAINTS OF THE JOURNALISTIC TRADE

1. The quote and the descriptive material are from interviews with Paul Jacobs, conducted by Professor Gerry Markle from Michigan State University, and are used with Markle's permission.

2. For discussion of these constraints as they affect general reporting, see Herbert J. Gans, *Deciding What's News* (New York: Random House, 1979), and Edward Jay Epstein, *News from Nowhere* (New York: Random House, 1973).

3. Time constraints may also affect reporting in different parts of the country. If a press conference is held at two o'clock and the deadline is five, writers will have little time to flesh out the story with background details. West Coast writers attending a scientific meeting on the East Coast will have the benefit of the time change. The timing of events with respect to deadlines can affect whether a topic gets covered and the thoroughness of the report.

4. *Communications and Medical Research Symposium Proceedings,* University of Pennsylvania, October 17, 1964, p. 32.

5. Sharon Dunwoody, "The Science Writing Inner Club," *Science, Technology and Human Values* 3, winter 1980, pp. 14–22.

6. Leo Bogart, "Editorial Ideals, Editorial Illusions," in Anthony Smith (ed.), *Newspapers and Democracy* (Cambridge, Mass.: MIT Press, 1980), pp. 247–278.

7. Kenneth Johnson, "Dimensions of Judgment of Science News Stories," *Journalism Quarterly* 40, 1963, pp. 315–322.

8. David Perlman, *NASW Newsletter,* December 1978.

9. Larry Sabato, *The Rise of Political Consultants* (New York: Basic Books, 1981).

10. *New York Times,* November 26, 1983.

11. Tom Johnson, "American Journalism: A Personal Perspective," *Phi Kappa Phi Journal,* winter 1983, pp. 29–30.

12. *New York Times,* November 26, 1983.

13. Similar avoidance of controversy is sometimes evident among the editors of scientific journals. For example, two NIH investigators had conducted a study of professional misconduct in science and found a high frequency of practices departing from accepted standards. Though the article had been reviewed by peers as an important public service, no journal editor was willing to publish their results. See correspondence from the files of Walter Stewart and Ned Feder of the National Institutes of Health during 1983 and 1984 covering their paper on "Professional Practices among Biomedical Scientists."

14. Oliver S. Moore III, quoted in Bernice Kanner, "Scientific Ex-

periments: Too Many Books?" *New York Magazine,* October 29, 1984, p. 16.

15. David Perlman, personal communication, January 24, 1984.

16. John Shelton Lawrence and Bernard Timberg, "News and Mythic Selectivity," *Journal of American Culture* 2, summer 1979, pp. 321–330.

17. Ben H. Bagdikian, "Conglomeration, Concentration and the Media," *Journal of Communication* 30, spring 1980, pp. 59–64.

18. Peter Dreier and Steve Weinberg, "Interlocking Directorates," *Columbia Journalism Review,* November/December 1979, pp. 51–68.

19. American Newspaper Publishers Association, *Facts about Newspapers* (Washington, D.C.: ANPA, 1982).

20. See discussion in John Westergard, "Power, Class and the Media," in James Curran, M. Gurevitch, and J. Woollacott (eds.), *Mass Communication and Society* (Beverly Hills, Calif.: Sage, 1979), pp. 95–115.

21. Quoted in Robert Cirino, *Don't Blame the People* (New York: Vintage Books, 1972), p. 93.

22. Alan Lupo, quoted in Dreier and Weinberg, op. cit.

23. See Seymour Hirsch, "On Uncovering the Great Nerve Gas Cover Up," *Ramparts* 3, July 1969, pp. 12–18.

24. Dwight Jensen, "The Loneliness of the Environmental Reporter," *Columbia Journalism Review,* January/February 1977, pp. 41–42.

25. This term was coined by Steven Hungerford and James B. Lemert, "Covering the Environment: New Afghanistanism?" *Journalism Quarterly* 50, autumn 1973, pp. 475–481, 508.

26. *New York Times,* February 3, 4, 5, and 6, 1980.

27. *New York Times,* February 7, 1980.

28. Telephone interview with Severo, November 1983. Also, see "Genetic Screening and the Handling of High-Risk Groups in the

Workplace," hearings before the Subcommittee on Investigations and Oversight of the Committee on Science and Technology, U.S. House of Representatives, 97th Congress, First Session, October 14, 15, 1981 (no. 53) (Washington, D.C.: Government Printing Office, 1982).

29. Jeffrey Kirsch, "On a Strategy for Using the Electronic Media to Improve Public Understanding of Science and Technology," *Science, Technology and Human Values* 4, spring 1979, pp. 52–58.

30. J. Lukasieweics, "The Ignorance Explosion," *Transactions* (New York Academy of Science), 1972, pp. 373–390. Also, Derek de Solla Price, *Science since Babylon* (New Haven: Yale University Press, 1961).

31. For reviews of accuracy in the press reports of complex scientific information see Sharon Dunwoody, "A Question of Accuracy," *IEEE Transactions on Professional Communication* PC–25, December 1982, pp. 196–199; James W. Tankard, Jr., and Michael Ryan, "Untangling the Numbers: Journalists Can Cope with Complex Research," *Newspaper Research Journal* 3, July 1982, pp. 61–69; and William Witt, "Effects of Quantification on Scientific Writing," *Journal of Communication* 1, winter 1976, pp. 67–69.

32. P. J. Tichener et al., "Mass Communication Systems and Communication Accuracy in Science News Reporting," *Journalism Quarterly* 47, winter 1970, pp. 673–683.

33. J. W. Tankard and M. Ryan, "News Source Perception of Accuracy of Science Coverage," *Journalism Quarterly* 51, summer 1974, pp. 219–334.

34. Dunwoody, op. cit.

35. See Sharon Friedman, "Environmental Reporting: Problem Child of the Media," *Environment* 25, December 1983, pp. 24–29; and Philip Tichenor, "Teaching and the Journalism of Uncertainty," *Journal of Environmental Education* 10, spring 1979, pp. 5–8.

36. For review of these technical uncertainties about the health effects of exposure to chemicals see introduction and technical appendix of Dorothy Nelkin and Michael S. Brown, *Workers at Risk* (Chicago: University of Chicago Press, 1984).

37.   Anthony Smith, *Goodly Gutenberg: The Newspaper Revolution of the 1980's* (New York: Oxford University Press, 1980), p. 188.

38.   David Blum, "A Kafkaesque Tale of Health Faddists Eating Cockroaches and Journalists Eating Crow," *Wall Street Journal*, September 28, 1981.

39.   Daniel Boorstin, *The Image* (New York: Harper & Row, 1964).

40.   Quoted in William Rivers, *The Adversaries: Politics and the Press* (Boston: Beacon Press, 1970), p. 49.

41.   Personal communication with Fred Golden of *Time*, November 1983.

42.   Quoted in Sharon Friedman, *Changes in Science Writing since 1965 and Their Relation to Shifting Public Attitudes toward Science*, master's thesis, Pennsylvania State University, 1974, p. 66.

43.   From discussions during a meeting of the Twentieth Century Fund Task Force on Risk and the Media, November 1983.

44.   David B. Sachsman, "Public Relations Influence on Coverage of Environment in San Francisco Area," *Journalism Quarterly* 53, spring 1976, pp. 54–60.

45.   Robert Gordon Shephard and Erica Goode, "Scientists in the Popular Press," *New Scientist* 76, November 24, 1977, pp. 482–484.

46.   Leon V. Sigal, *Reporters and Officials* (Lexington, Mass.: D. C. Heath, 1973).

47.   Mary Marlino, *Reporting on PCB's in the Hudson*, master's thesis, Cornell University, 1984.

48.   Nancy Hicks, statement in discussion during a symposium on medicine and the media, University of Rochester Medical Center, October 9–10, 1975, *Proceedings*, p. 133.

Chapter 8. THE PUBLIC RELATIONS OF SCIENCE

1. Quoted in the *New York Times,* March 16, 1985.

2. For discussion of the norms and communication practices of science, see Robert Merton, *Sociology of Science* (New York: University of Chicago Press, 1973), and Bernard Barber and Walter Hirsch (eds.), *The Sociology of Science* (New York: Free Press, 1962).

3. Quoted in Hillier Krieghbaum, "American Newspaper Reporting of News," *Kansas State Bulletin,* August 15, 1941, p. 20.

4. Ronald Tobey, *The American Ideology of National Science 1919– 1930* (Pittsburgh: University of Pittsburgh Press, 1971).

5. *NASW Newsletter,* June 1, 1954, p. 10.

6. George Kistiakowsky, speech to the National Association of Science Writers, quoted in "Our Twenty-Fifth Anniversary," *NASW Newsletter* 7, December 1959, pp. 11–12.

7. Pat McGrady, public relations officer of the American Medical Association, in a 1967 review, quoted in Carolyn Hay, *A History of Science Writing in the United States,* master's thesis, Northwestern University, 1970, p. 254.

8. Jean Rostand, "Popularization of Science," *Science* 131, May 1960, p. 1491.

9. Quoted in Gerard Piel, "Democracy and the Intellect," in Miriam Balaban (ed.), *Science Information Transfer: The Editor's Role* (Amsterdam: Elsevier, 1978).

10. Robert H. Grant and Kenneth D. Fisher, "Scientists and Science Writers: Concerns and Proposed Solutions," *FASEB Proceedings* 39, 1971, pp. 816–826.

11. *New York Times,* April 25, 1986.

12. Edward Feigenbaum and Pamela McCorduck, *The Fifth Generation* (Reading, Mass.: Addison-Wesley, 1983), p. 8.

13. See Wilson interview in *SIPIScope* 13, November/December 1985, p. 11.

14. Arthur Caplan, "So How Are We Doing?" paper presented at the annual meeting of the American Society of News Editors, April 10, 1985.

15. Phil Gunby, "Media-Abetted Liver Transplants Raise Questions of Equity and Decency," *Journal of the American Medical Association* 249, April 1983, pp. 1973–1982.

16. *Courier Journal*, November 27, 1984.

17. *Courier Journal*, March 3, 1985.

18. Robert S. Cowen, "Garbage under Glass: What Are Scientists Dishing Out?" *Technology Review* 81, November 1979, pp. 10–11.

19. *NASW Newsletter*, October 1981, p. 12.

20. Jay Winsten, "Science and the Media: The Boundaries of Truth," *Health Affairs* 4, spring 1985, pp. 5–23.

21. William Allman, "The Lost Story of Peru," *Science 85*, May 1985, p. 16.

22. Many industrial scientists are willing to play this role. Studies suggest that while corporate scientists may face conflicts between their expectations of scientific autonomy and the pragmatic demands of bureaucratic loyalty, they tend nonetheless to adopt managerial values, which include concerns about enhancing corporate image. This loyalty guides their response to the press. For studies of industrial science see William Kornhauser, *Scientists in Industry* (Berkeley: University of California Press, 1962); Barney Glaser, *Organizational Scientists* (Indianapolis: Bobbs Merrill, 1964); R. Ritti, *The Engineer in the Industrial Corporation* (New York: Columbia University Press, 1971); and Edwin A. Layton, *The Revolt of the Engineers* (Cleveland: Case Western Reserve Press, 1971).

23. Richard Tucker, lecture to Clinical Institute of Toxicology, 1983 (mimeo).

24. Cited in Stephen Hilgartner, Richard Bell, and Rory O'Connor, *Nukespeak* (San Francisco: Sierra Club Books, 1982), p. 77.

25. Ibid., p. 78.

26. Ibid., p. 79.

27. Michael D. Tabris, "Surviving a State of Siege," *The Chemist*, May 1983, pp. 5, 22.

28. Richard Tucker, op. cit.

29. Information excerpted from their public relations material. In 1982, the scientists' lectures were reported in newspapers in Atlanta, San Bernardino, Houston, Phoenix, Oklahoma City, and Charleston.

30. These phrases are from public relations brochures.

31. *Wall Street Journal*, July 22, 1982, p. 23.

32. Francesca Lunzer, "Medicine by Press Agentry," *Forbes*, September 24, 1984, pp. 199–202.

33. See *Science* 224, June 1984, pp. 1392–1398. Critiques are reviewed in *Medical World News*, August 27, 1984, pp. 38–46 and *Science* 225, August 1984, pp. 705–706. See also the *Washington Post*, June 23, 1984.

34. I. Stanton, "Public Relations and the Press," in Gary Wagner (ed.), *Publicity Forum* (New York: R. Weiner, 1977), p. 31.

35. "Declarations of Principles," published in William L. Rivers, Wilbur Schram, and Clifford Christian, *Responsibility in Mass Communications* (New York: Harper & Row, 1980), Appendix E.

36. *NASW Newsletter, passim*, 1979.

37. David Zimmerman, "How Consumer Education, Public Relations Style, Still Translates as 'Press Agentry'," *NASW Newsletter*, April 1979.

38. Quoted in Michael Schudson, *Discovering the News* (New York: Basic Books, 1978), p. 139.

39. Quoted in Dan Schiller, *Objectivity and the News* (Philadelphia: University of Pennsylvania Press, 1981), p. 188.

40. These statements are from letters and discussion in the *NASW Newsletter*, from which innumerable other examples could also be drawn. See, for instance, Mark Bloom, "The Editor's Letter," *NASW Newsletter*, April 1979, and letters to the editor, passim.

Chapter 9. HOW SCIENTISTS CONTROL THE NEWS

1. Personal discussion with a visiting speaker at the NASW New Horizons Meeting, Blacksburg, Virginia, November 1983.

2. Philip Handler, "Public Doubts about Science," *Science* 108, June 6, 1980 (editorial).

3. George Keyworth was interviewed by Fred Jerome in *SIPIScope*, February 1985.

4. "The Poisoning of America or . . . the 'Poisoning' of America," in *The Point Is . . .* (Dow publication), October 7, 1980.

5. Harold Schmeck's article appeared in the *New York Times* on May 27, 1980; the reply appeared on June 17, 1980.

6. Arnold Relman comment in "Medicine in the Media: Panel Discussion," *P & S* 2, April 2, 1982, p. 16.

7. *New York Times*, March 16, 1986.

8. Quoted in Lawrence K. Altman, "Publicity and Medicine," *New York Times*, January 1, 1985.

9. Arthur Sackler, at symposium on medicine and the media, University of Rochester Medical Center, October 9–10, 1975, *Proceedings*, p. 86.

10. For discussion of the norms of open communication see "The Normative Structure of Science," in Robert K. Merton, *The Sociology of Science* (Chicago: Chicago University Press, 1973), pp. 267–278. Also see Sissela Bok, "Secrecy and Openness in Science," *Science Technology and Human Values* 7, winter 1982, pp. 32–41.

11. For discussion of the changing pressures on science and their effect on open communication see Dorothy Nelkin, *Science as Intellectual Property* (New York: Macmillan, 1984).

12. Sharon Dunwoody and Michael Ryan, "Scientific Barriers to the Popularization of Science—The Mass Media," *Journal of Communication* 35, winter 1985, pp. 26–42.

13. A much-quoted repartee. See David W. Burkett, *Writing Science News for the Mass Media* (Houston: Gulf, 1973), p. 132.

14. Martin Bander, "The Scientist and the News Media," *New England Journal of Medicine* 308, May 1983, pp. 1170–1173. See also Neal E. Miller, "The Scientist's Responsibility for Public Information: A Guide to Effective Communication with the Media," *Society for Neuroscience*, (Bethesda, Maryland: 1978).

15. Rae Goodell, *The Visible Scientists* (Boston: Little Brown, 1977), p. 61.

16. Sharon Dunwoody and Byron T. Scott, "Scientists as Mass Media Sources," *Journalism Quarterly* 59, spring 1982, pp. 52–59; Luc Boltanski and Pascale Maldidier, "Carriere Scientifique, Morale Scientifique et Vulgarisation," *Information Science and Society* 9 (3), 1970, pp. 99–118.

17. Franz J. Ingelfinger, "Medical Literature: The Campus without Tumult," *Science* 169, August 1970, p. 733.

18. Relman has written and lectured extensively on his policy. See *New England Journal of Medicine* 303, December 1980, pp. 1527–1528; *Bryn Mawr Alumni Bulletin*, fall 1981, pp. 2–5; *NASW Newsletter*, November 1979, pp. 9–10; and "Special Report on Medicine and the Media," *P & S* 2, April 1982, p. 16.

19. Quoted in June Goodfield, *Reflections on Science and the Media* (Washington, D.C.: American Association for the Advancement of Science, 1981), p. 94.

20. Symposium on medicine and the media, University of Rochester Medical Center, October 9–10, 1975, *Proceedings*, p. 182.

21. Alan Westin, *Whistleblowing: Loyalty and Dissent in the Corporation* (New York: McGraw-Hill, 1981).

22. Nelkin, op. cit.

23. Quoted in Mary Marlino, *Reporting on PCB's in the Hudson,* master's thesis, Cornell University, 1984, p. 38.

24. See discussion in President's Commission on the Accident at Three Mile Island, *Report of the Public Right to Information Task Force* (Washington, D.C.: Government Printing Office, 1979).

25. *New York Times,* May 22, 1986.

26. Michael Altimore, "The Social Construction of a Scientific Controversy: Comments on Press Coverage of the Recombinant DNA Debate," *Science, Technology and Human Values* 7, fall 1982, pp. 24–31.

27. Symposium on medicine and the media, University of Rochester Medical Center, October 9–10, 1975, *Proceedings*.

28. *Philadelphia Inquirer*, September 25, 1983.

29. Louis Lasagna, symposium on medicine and the media, University of Rochester Medical Center, October 9–10, 1975, *Proceedings*, p. 7.

30. Ibid., p. 58.

31. Barry R. Bloom, "News about Carcinogens: What's Fit to Print," *Hastings Center Report*, August 1979, pp. 5–7.

32. *NASW Newsletter*, April 1, 1982.

33. Symposium on medicine and the media, University of Rochester Medical Center, October 9–10, 1975, *Proceedings*, p. 34.

34. Reported in William Colglazier, Jr., and Michael Rice, "Media Coverage of Complex Technological Issues," in Dorothy Zinberg (ed.), *Uncertain Power* (New York: Pergamon, 1983), p. 113.

35. G. Bugliarello, "A Technological Magistrature," *Bulletin of the Atomic Scientists* 1, 1978, pp. 34–37; and J. C. White, "Reflections and Speculations on the Revelation of Molecular Genetic Research," *Annals of the New York Academy of Sciences* 265, 1976, p. 173.

36. Sharon M. Friedman, "Blueprint for Breakdown: Three Mile Island and the Media before the Accident," *Journal of Communication* 31, spring 1981, pp. 116–128.

37. Frank Graham, Jr., *Since Silent Spring* (Boston: Houghton Mifflin, 1970), pp. 165–166.

38. *New York Times*, February 20, 1983.

Chapter 10. THE HIGH COST OF HYPE

1. This section on the Challenger accident was developed with the help of Susan Lindee, a Cornell University doctoral student.

2. *Houston Chronicle*, February 1, 1986; *Miami Herald*, February 23, 1986; and *New York Times*, February 5, 1986.

3. *New York Times*, March 20, 1986.

4. For a theoretical perspective on the hegemonic role of the press see Stuart Hall, "Culture, the Media and the Ideological Effect," in James Curran, M. Gurevitch, and J. Woollacott (eds.), *Mass Communication and Society* (Beverly Hills, Calif.: Sage, 1979), chap. 13.

5. See discussion in Robert Young, "Science as Culture," *Quarto*, December 1979, p. 7; and Langdon Winner, "Mythinformation," in Paul T. Durbin (ed.), *Research in Philosophy and Technology*, vol. 7 (Greenwich, Conn.: JAI Press, 1984), pp. 287-304.

6. There are, of course, notable exceptions. Daniel Greenberg, who publishes a newsletter entitled *Government and Science Report* and sometimes writes a science policy column for the *Washington Post*. The *New York Times* has hired several reporters from *Science* who have been writing some interpretive and critical articles. The news and comments section of *Science* also publishes critical commentary; however, *Science* is a specialized publication read mainly by scientists and science policy professionals.

7. L. E. Trachtman, "The Public Understanding of Science Effort," *Science, Technology and Human Values*, vol. 6, summer 1981, pp. 10-15.

8. Arnold Relman, "Special Report on Medicine and the Media," *P & S*, April 1982, p. 22.

9. See Erving Goffman, "Felicity's Condition," *American Journal of Sociology* 89, July 1983, pp. 1-53.

10. J. C. Pocock, "Ritual Language and Power," *Politics, Language and Times* (London: Methuen, 1970).

11. Richard Whitley, in Terry Shinn and Richard Whitley (eds.), *Expository Science* (Dordrecht: D. Reidel, 1985), points out that the many different nonscientific groups that constitute the audience for scientific information seek such information for specific purposes and assimilate it accordingly.

12. Quoted in Harold I. Sharlin, *EDB: A Case Study in the Communication of Health Risk*, (Washington, D.C.: Environmental Protection Agency, 1985).

13. See, for example, David Rubin, "What the President's Commission Learned about the Media," in T. Moss and D. Sills (eds.), *The Three Mile Island Accident* (New York: New York Academy of Sciences, 1981), pp. 95–106.

14. Howard Blakeslee, (Associated Press), Ferry Colton (Associated Press), Watson Davis (Science Service), David Dietz (Scripps Howard), Victor Henderson (*Philadelphia Inquirer*), Thomas Henry (*Washington Star*), Waldemar Kaempffert (*New York Times*), Gobind Behari Lal (Hearst), William Laurence (*New York Times*), John O'Neill (*New York Herald Tribune*), Robert B. Potter (*American Weekly*), and Allen Shoenfield (*Detroit News*).

15. The charter is reproduced in Carolyn D. Hay, *A History of Science Writing in the United States*, master's thesis, Northwestern University, 1970.

16. Quoted in Hay, op. cit., p. 54.

17. Another professional newsletter, *Sciphers*, appeared in 1979, published by the Science Writing Educators Group at the University of Missouri.

18. A similar organization, the Center for Health Communication, has been formed at the Harvard School of Public Health to clarify and interpret health information for journalists. The communications office of the American Association for the Advancement of Science plays a similar role.

I N D E X

Accuracy, 125–126, 156
Advertising
  "advertorials," 144–145
  high technology in, 37
  newspaper revenue
    from, 121
  science manipulated in,
    79, 151
  on television, 124
Afghanistan syndrome, 122
Agassiz, Louis, 86, 134
AIDS (acquired immune
    deficiency syndrome),
    3, 52, 83, 110–111, 120,

143, 173
Allen, Frederick Lewis, 87
Alsop, Joseph, 128
Altman, Lawrence, 43,
    110–111, 113
Alzheimer's disease, 142
American Association for the
    Advancement of Science,
    88, 212 *n*. 18
American Cancer Society
    (ACS), 4, 105,
    129, 135
American Chemical Society,
    55, 134

American Institute of
Physics, 137
American Medical
Association, 48, 134
American Physical
Society, 167
American Society for the
Control of Cancer,
134–135
Animal tests for
carcinogenicity, 62–63
Anorexia nervosa, 150
Arkansas Gazette, 67
Arthritis, 149–150
Artificial hearts, 42–45,
112, 157
public relations for,
139–141
Artificial sweeteners dispute,
64–68, 82, 144, 178–179
Asbestosis, 104
Asilomar conference (1975),
39–40, 165
Associated Press (AP), 9, 92
on cancer research, 105
corrections issued by, 142
on estrogen replacement
therapy, 45, 46
on ozone controversy, 56
Athenaeum, 86
Atlanta Journal, 66
Audience, journalists' and
scientists' perceptions of,
118–120, 177–178
Automobile industry, 37
Autonomy
of industrial scientists,
206 *n.* 22
of science writers, 116
Ayerst Laboratories
(firm), 148

Baby Fae experiment, 140
Barnard, Christiaan,
41–42, 151
Basalla, George, 75

Benbow, Camille, 29–30
Bennett, James Gordon, 93
Bethe, Hans, 18
Biases
of public, 77
in public policy, 80–81
of science writers, 68,
100–108, 192 *n.* 28
Biogen (firm), 4, 5
Biological determinism, 27–31
Biotechnology, 38–41, 105,
118, 142
Birth control, 47, 79
Blakeslee, Alton, 105
Bloom, Barry R., 166
Boston Globe, 25, 28, 171
Boyle, Joseph, 157
Brodeur, Paul, 103
Brody, Jane, 47
Bronowski, Jacob, 136
Brown, Louise, 50–51
Burke, Kenneth, 10
Burnham, David, 103–104
Burrows, William, 8
Business reporting
in artificial sweeteners
dispute, 60
on computers, 35
on interferon, 5
on science education, 22–24
on sociobiology, 30
on toxic wastes, 67–68
Business Week, 9
on artificial sweeteners
dispute, 59, 60
on ozone controversy, 56
on science education, 23
on sociobiology, 30

Calcium, 150–151
Caldicott, Helen, 149
Calorie Control Council,
62–63, 144, 148
Cancer
artificial sweeteners and,

59, 62–63
Delaney Amendment
    and, 58
estrogen replacement
    therapy and, 47
interferon reporting and,
    4–6
laetrile and, 82
politics of research on, 105
public interest in, 77
public relations for research
    in, 129
publication of data on, 166
smoking and, 80
Carcinogenicity, animal tests
    for, 62–63
Carey, Frank, 3
Center for Health
    Communications,
    212 n. 18
Challenger (space shuttle), 3,
    120, 137, 157, 170–172
Chemical industry, 103
advertising by, 145
genetic screening of
    workers in, 123–124
press blamed by, 155–156
public relations by, 146–147
Chernobyl nuclear accident,
    71, 164–165, 167
Chlorofluorocarbons, 55–58
Christian Science Monitor, 25,
    36, 141
Clark, Barney, 42–43, 139–140
Cline, Martin, 109–112
"Clockwork Orange"
    (film), 49
Cobb, Frank, 152
Cohen, Morris, 88
Cohn, Victor, 119, 163
Colorado, University of, 143
Communications research,
    73–77
Competition
    in coverage of Nobel prize
    winners, 15–16

for research funding, 158
as theme in science
    journalism, 7, 35
Computers, 22–23, 99,
    113, 118
promoted by press, 34–37
public relations for, 138
supercomputers, 38,
    132–133
Contraceptives, 47, 79
Cooley, Charles, 73
Cooley, Denton, 42
Corporations
influences on editorial
    policy by, 122–124
public relations by, 144–153
scientists in, 163
Cosmopolitan, 28, 29
Council for the Advancement
    of Science Writing
    (CASW), 181
Cowen, Robert C., 141
Cronin, James, 17
Cyclamates, 60, 82

Dapsone (drug), 166
Dartmouth Medical
    Center, 142
Deadlines, 113–114,
    200–201 n. 3
DeBakey, Michael, 42
Delaney Amendment to the
    Food, Drug, and Cosmetic
    Act, 58, 60, 64
DeVries, William, 140
Dietz, David, 180
Dioxin, 64–68, 127, 169, 178
Discover, 9, 30
Dow Chemical, 66, 67,
    147, 155
Drugs, public relations for,
    149–150
Dunwoody, Sharon, 113
Du Pont (firm), 56, 57,
    123–124, 144

Economic issues
in industrial public
relations, 144–153
effect on science writers,
105, 120–124
Edelin, Kenneth, 83–84
Edelson, Ed, 107
Editorial constraints on
science journalism,
114–118
Edwards, Charles, 60
Electric power industry,
145–146. *See also* Nuclear
energy
Environmental Protection
Agency (EPA), 58
on toxic wastes, 65, 67,
68, 82
Environmental reporting,
84, 97
chemical workers not
covered in, 103–104
control of information for,
168–169
on dioxin, 64–68
economic constraints
on, 122
governmental secrecy
and, 164
press releases and, 129
sources used for, 130–131
Epidemics, 178
Epstein, Samuel, 60
Errors, in science articles, 126
Estrogen replacement therapy
(ERT), 45–47, 148
Ethics
of journalists, 91, 105
of public relations
professionals, 151
Eugenics, 90
Evidence, 178

Family Health, 19
Family Week, 28
Feigenbaum, Edward, 138

Fetal research, 83
Fiske, Jamie, 139
Fiske, Richard, 166–167
Fitch, Val, 17
Fluorocarbons, 55–58
Food additives, 58–64
Food and Drug
Administration (FDA)
in artificial sweeteners
dispute, 58, 60, 64,
82
on Oraflex and
Virazole, 150
Fraud in science, 24–26, 84
Freeman, Walter, 48
Funding
competition in science
for, 158
for health-related
programs, 83
public relations for,
136–137, 141

Gans, Herbert J., 104,
197–198 *n.* 18
Genetic engineering, 109–110
Genetics
recombinant DNA research,
39–41
in sociobiology, 27–31
Genetic screening of workers,
123–124
Gerbner, George, 74
Germany, 87
Gilbert, Walter, 18
Gitlin, Todd, 9, 95
Gofman, John, 149
Goodell, Rae, 160–161
Gould, Stephen J., 31
Government agencies
censorship and restrictions
by, 163–164
public relations by, 137
as sources, 130
Government and Science Report
(newsletter), 211 *n.* 6

Graham, Frank, Jr., 168
Gray, Asa, 86
Greenberg, Daniel, 61,
    211 *n.* 6

Handler, Philip, 155
Handlin, Oscar, 78
Harvard Medical School, 142
Harvard University, 39
Headlines, 115
Health. *See* Medical reporting
Heart transplantation, 3,
    41–42, 112
  artificial hearts for, 42–45,
    139–141, 157
Henry, Joseph, 134
Herschel, John, 86
Hess, Stephen, 184–185 *n.* 9
High technology, 34–35
  medical applications of,
    41–51
  promotion of, 10, 35–41
  science literacy for, 21–22
  See also Technology
Hill and Knowlton (firm),
    47, 148
Hoaxes, 128
Hodgkin, Dorothy, 19
Hoffmann, Roald, 16, 17
Hooker Chemical Company,
    64, 103
Hospitals, public relations by,
    139–140
Hounsfield, Sir Godfrey, 17
Hudson, Rock, 83
Humana Hospital, 44,
    140–141, 157
Human subjects, genetic
    engineering on, 109–110
Huxley, Aldous, 50
Huxley, Thomas Henry, 14,
    86, 134

Industrial and Scientific
    Communications Service
    (ISCS), 148

Industrial public relations,
    144–153
Ingelfinger, Franz J., 161, 162
Institutional Review Boards,
    43, 44, 110
Interferon, 4–7, 83, 156
In vitro fertilization (IVF),
    50–51
Isselbacher, Kurt, 60

Jacobs, Paul, 109–111
Japan, 35–36
Jensen, Arthur, 27
Johns Manville
    Corporation, 103
Johnson, Mark, 10–11
Journalism
  during Nixon
    administration, 116–117
  norms of objectivity and
    fairness in, 54
  public policy issues
    "framed" by, 80–84
  scientists on role of, 8, 179
  Upton Sinclair on, 152
  tension between science
    and, 182
*Journal of the American Medical
    Association,* 46, 162
Journals (scientific)
  guidelines on scientist-
    journalist contacts
    published in, 160
  "information explosion"
    in, 125
  "Ingelfinger rule" in,
    161–162
  public relations by, 137–138

Kaempffert, Waldemer, 91
Kansas City Star, 36
Keyworth, George A., II, 155
Kirsch, Jeffrey, 124
Kistiakowsky, George, 136
Kistner, Robert A., 46–47

Klapper, Joseph, 74
Kornberg, Arthur, 106
Krim, Mathilde, 4
Kubrick, Stanley, 49

Laetrile, 82
LaFollette, Marcel, 75
Lakoff, George, 10-11
Lal, Gobind, 91
Lamb, Larry, 165-166
Language
    in high-technology
        advertising, 37
    in high-technology
        articles, 35
    of metaphors in science
        journalism, 10-11, 61, 81
    of science journalism,
        changes in, 9-11, 99-100
    as used by journalists and
        scientists, 177-179
    used in public
        relations, 146
Lasagna, Louis, 165
Laski, Harold, 81
Laurence, William, 1, 2, 91
Lawsuits against
    journalists, 117
Lazarsfeld, Paul, 73
Lear, John, 97
Lester, Marilyn, 95
Lewontin, Richard, 31
Life, 41
Lilly (firm), 149-150
Lippmann, Walter, 73
Lobotomy (psychosurgery),
    48-50
Look, 46
Los Angeles Times, 66, 109-111
Love Canal (N.Y.), 64-65,
    82, 127

McAuliffe, Christa, 171
McCarron, David, 150-151
McClintock, Barbara, 19-20
McElheny, Victor, 106-107

McLuhan, Marshall, 74
Marijuana research, 130
Mathematics, sex differences
    in, 29-30, 141-142
Mayer, Maria, 18-19
McCall's, 18-19, 47
Media
    public attitudes influenced
        by, 72-80
    public policy issues
        "framed" by, 80-84
Media consultants, 117
Media Resource Service
    (MRS), 181-182
Medical reporting
    accuracy and error in, 126
    on AIDS, 52
    on artificial hearts and
        transplants, 41-45
    on artificial sweeteners
        dispute, 58-64
    changes in, 98-99
    "Cohn's First Law" of, 119
    corporate influences on,
        123-124
    on dioxin, 64-68
    on estrogen replacement
        therapy, 45-47
    evidence in, 178
    on genetic engineering,
        109-110
    on interferon, 4-7
    on in vitro fertilization,
        50-51
    occupational health avoided
        in, 103-104
    on psychosurgery, 48-50
    public interest in, 77
    public relations and,
        139-143, 149-151
    restrictions proposed for,
        165-167
Medical schools, public
    relations by, 139
Menopause, estrogen
    replacement therapy for,
    45-47

Merck (firm), 149
Metaphors, 10–11, 81,
    185 n. 15
Metropolitan Edison, 144, 148
Miami Herald, 170
Miller, Jonathan, 75
Milwaukee Journal, 22, 66
Molina, Mario J., 55
Molotch, Harvey, 95
Museums, 183 n. 1
Mystification of science,
    17–18

Nation, 14
National Academy of Sciences
    (NAS)
  on artificial sweeteners
    dispute, 60–62, 178–179
  on ozone controversy, 57
  public relations by, 137
  Science Service and, 88
National Aeronautics and
    Space Administration
    (NASA), 168
  Challenger accident and,
    137, 157, 170–172
National Association of
    Science Writers (NASW),
    152–153, 180–181
National Cancer Institute
    (NCI), 166
National Dairy Council, 151
National Enquirer, 76
National Institutes of
    Health, 83
National Science
    Foundation, 132
Nature (journal), 105
New England Journal of
    Medicine, 105,
    137–138, 165
  on estrogen replacement
    therapy, 47
  guidelines for scientists
    in, 160
  "Ingelfinger rule" in,
    161–162

New Haven Register, 56
New Republic, 94
Newsbank, 40
Newspapers
  constraints on writers for,
    91–92
  economic pressures on,
    120–124
  editorial constraints in,
    114–118
  history of objectivity in,
    92–95
  influence of, 8–9
  news value of science in,
    111–112
  in nineteenth century,
    85–87
  read by reporters,
    184–185 n. 9
  tabloids, 76
Newsweek, 9, 35, 112
  on artificial sweeteners
    dispute, 59, 63
  on Challenger accident, 171
  on competition with
    Japan, 36
  on fraud in science, 26
  on high technology, 37–38
  on interferon, 5
  on in vitro fertilization, 50
  on Nobel prize winners,
    19–20
  on sociobiology, 28, 31
New York Herald, 93
New York Times, 2, 8–9, 113,
    125, 130–131, 142,
    184–185 n. 9, 211 n. 6
  on AIDS, 110–111
  on artificial heart, 43, 44
  on artificial sweeteners
    dispute, 61
  on Challenger accident,
    170–171
  on Cline case, 111–112
  on Dapsone, 166
  on estrogen replacement
    therapy, 47

New York Times (continued)
on fraud in science, 25
on genetic screenings of
workers, 123–124
on high technology, 38
on interferon, 5
on in vitro fertilization, 50
on Nobel prize winners,
18–20
on ozone controversy,
55, 57
on psychosurgery, 49–50
on public understanding of
science (1919), 88
on recombinant DNA
research, 39–40
on science education, 21
on sociobiology, 28
on toxic wastes, 67–68
New York Tribune, 86
Niagara Gazette, 64
Nixon administration,
116–117
Nobel prize winners,
15–21, 176
Nuclear energy, 99, 120
Chernobyl nuclear accident
and, 71
impact of press coverage
on, 72
official accounts of, 167–168
public relations for,
144–146
secrecy over accidents in,
164–165
Three Mile Island and,
80, 148

Objectivity
by journalists and by
scientists, 177
as norm in journalism, 54,
91–96
in science, questioning
of, 60

Occidental Chemicals (firm),
146–147
Occupational health, 103–104
Omenn, Gilbert, 78
Oraflex (drug), 149–150
Oral contraceptives, 47
Ozone controversy, 55–58, 77

PBB, 168–169
PCBs, 130
Peer review, 161, 175
Perlman, David, 97–101,
105–106, 113, 115, 119
Pesticides, 99, 168–169
Philadelphia Inquirer, 55–56
Philadelphia Public Ledger, 93
Pierce, Henry, 96–97
Pittsburgh Post Gazette, 45
Placement of articles, 115
Playboy, 28, 29
Polio vaccine, 79–80
Political reporting, 104–107,
174, 175
science reporting compared
with, 116–117
Politics, science separated
from, 105
Popularization of science,
74–75, 136
Press conferences, 129–130
Press releases, 128–129. See
also Public relations
Professional associations, 134
Pseudoscience, 75–76
Psychology Today, 28
Psychosurgery, 48–52, 54
Public health
control of information
on, 169
prepublication release of
news related to, 162
Public opinion
media impact upon, 72–80
on press reliability, 117

Public policy
impact of science writers
on, 172
issues "framed" by media
in, 9–10, 80–84
Public relations, 132–133
in artificial heart implants,
43–45
in artificial sweeteners
dispute, 62–63
in Challenger accident, 171
in dioxin disputes, 67
for estrogen replacement
therapy, 46, 47
by government agencies,
scientific and technical
institutions, 117
industrial, 144–153
in interferon coverage, 4
in ozone controversy, 56
press releases and, 128–129
for psychosurgery, 48
for scientific institutions,
134–143
by scientists, 7–8, 13, 160

Readers, assumptions about,
118–120
Reader's Digest, 5, 165
Reader's Guide, 48–50
Recombinant DNA research,
39–41, 82, 165
human experimentation in,
109–110
public relations for, 138
Relman, Arnold, 156, 157,
161–162, 176
Right to know, 159
Risk reporting, 3, 54, 68–69
accuracy and error in,
126–127
on artificial sweeteners
dispute, 58–64
changes in, 99–100
on dioxin, 64–68

estrogen replacement
therapy and, 46–47
ideological assumptions
in, 71
language used in, 119
objectivity norm in, 92
on recombinant DNA
research, 39
secrecy in, 165
social biases in, 105
Ritter, William E., 88, 89
Robbins, Fred, 62
Roberts, Wade, 104
Rostand, Jean, 136
Rothmeyer, Karen, 104
Rowland, F. Sherwood, 55, 57
Rubin, David, 184 n. 2

Saccharin, 58, 59, 61–63, 126,
148, 178–179
Sachsman, David, 129
Sagan, Carl, 74–75
Salk polio vaccine, 79–80
Salt Lake Tribune, 36, 44
Salton, Lawrence, 134
San Francisco Chronicle, 113
San Jose Mercury, 22
Saturday Evening Post, 5, 87
Schiller, Dan, 92
Schmeck, Harold, 5, 156,
184 n. 2
Schroeder, William, 44–45,
112, 140–141
Schwartz, Harry, 44
Science, 2, 14–21
critical reporting on, 175,
211 n. 6
fraud in, 24–26, 84
ideals of, accepted by
journalism, 71–72, 92–94
politics separated from, 105
public interest in, 76–77, 79
public perceptions of, 79
as resource, 21–24,
59–60, 66

Science (*continued*)
  on television, 74–75
  training in, for journalists,
    102–103, 181
Science (journal), 14, 29–30,
    105, 138, 150–151, 211 *n*. 6
Science 86 (journal), 121
Science Digest, 9, 19, 28, 31,
    118, 121
Science education, 15,
    21–24, 81
  computer education in, 36
Science journalism
  changes in (1960s–1980s),
    10, 96–100
  complexity constraints on,
    124–128
  critical, 175, 211 *n*. 6
  economic pressures in,
    120–124
  editorial constraints on,
    114–118
  general characteristics of,
    6–9, 11–12
  goals of, 180
  history of, 85–90,
    196–197 *n*. 2
  homogeneity of, 9
  ideological assumptions in,
    71, 100–108
  influence of, 72–84
  language used in, 10–11
  newswork in, 111–114
  objectivity as norm in,
    91–96
  public's biases and, 77
  scientists' control over,
    159–169
  scientists on role of,
    155–159, 179
  social biases in, 100–108
  sources for, 128–131
Science literacy. *See* Science
    education
Science museums, 183 *n*. 1
Science News, 31

Science Service, 88–90, 134
Science writers
  autonomy of, 116
  in early twentieth century,
    90–91
  impact on public policy
    of, 172
  manipulation of, 174
  press releases used by,
    128–129
  professional associations of,
    180–182
  public relations pressures
    on, 47–48, 152–153
  self-censorship of, 116–118
  social biases of, 100–108
Scientific knowledge
  control over, 158–159
  increases in, 125
Scientific societies, 134
Scientists
  criticism of press, 126,
    155–159
  image in press of, 15–21
  in industrial public
    relations, involvement of,
    144–153
  Media Resource Service
    lists of, 181–182
  news controlled by,
    159–169
  portrayed on television,
    74–75
  public perceptions of,
    78, 88
  public relations by, 13,
    134–138, 141–142
  relationships between
    journalists and, 8,
    130–131, 155–159, 182
Scientists' Institute for Public
    Information (SIPI), 181
Scripps, Edwin W., 88–89, 134
Seaborg, Glenn, 22
Secrecy
  corporate, 163

by governments, 164–165
in scientific
communications, 158
Self-censorship of science
writers, 116–118
Selikoff, Irving, 67
Severo, Richard, 123–124
Sex differences
in mathematical ability,
29–30, 141–142
sociobiology on, 28–29
Sigel, Leon, 130
Sigma Delta Chi (Society of
Professional Journalists),
91, 92
Sinclair, Upton, 152
Slosson, Edwin E., 89–90
Smith, Anthony, 128
Smith, Dorothy, 55
Smoking, 80
Social class biases, 104–105
Social policy, science in
formation of, 26
Society of Professional
Journalists (Sigma Delta
Chi), 91, 92
Sociobiology, 27–31
Sources for stories, 130–131
public relations officers
as, 141
Space exploration, 34,
106–107, 157
Challenger accident and,
170–172
control of news on, 168
public relations for, 137
Sports reporting, 106, 107
Sputnik, 21, 135–136
Stanford University, 139
Stanley, Julian, 29–30
Star, 76
Stockton, William, 102
Straus, Marc J., 25
Stroube, Hal, 145–146
Sullivan, Walter, 55, 101, 125,
184 n. 2

Sunday World, 85
Surrogate motherhood, 51
Sweeteners, artificial, 58–64

Tabloid newspapers, 76
Tabris, Michael, 146–147
Technology
accuracy and error in
reporting of, 126, 156
Challenger accident and,
170–171
changes in reporting of,
99–100
coverage of, 2, 33–34,
72–73, 173–175
high technology, 34–35
language used in reporting
on, 10–11, 119
medical applications of,
41–51
in nineteenth-century
journalism, 86
promotion of, 10, 35–41
public interest in, 76–77
reporting on risks in, 3,
53–54
science literacy for, 21–22
Television, 183 n. 1
economic pressures on, 124
newspaper formats tied
to, 120
science on, 74–75
Three Mile Island, 80, 144,
148, 164, 168
Time, 9
on artificial sweeteners
dispute, 60
on estrogen replacement
therapy, 46
on interferon, 5
on in vitro fertilization, 50
on Nobel prize winners, 16,
17, 19, 20
on recombinant DNA
research, 39

Time (continued)
  on science education, 23
  on sociobiology, 28, 29
  on toxic chemicals, 65–67
Times Beach (Missouri),
  65, 169
Toxic shock syndrome, 79
Toxic substances, 127, 178
Toxic wastes, 80, 82, 103, 156
  control of information
    on, 169
  corporate public relations
    and, 147
  dioxin, 64–68
Tuchman, Gaye, 95
Tucker, Richard, 144, 147
Twain, Mark, 94
Tyndall, John, 86, 134

Ubell, Earl, 113, 163
Uncertainty, 115–116, 127,
  166–167
Union of Concerned
  Scientists, 149
United Press International
  (UPI), 9, 65, 128
Universities, public relations
  by, 138–139
U.S.A. Today, 120
U.S. News and World Report, 9
  on Nobel prize winners,
    15, 21
  on technological change, 33
Utah Medical Center, 42–44,
  139–140

Virazole (drug), 150
Vogue, 19, 47

Wall Street Journal, 8–9, 56,
  185 n. 9
Washington Post, 8–9, 92, 130
  on artificial sweeteners
    dispute, 61–62
  on fetal research
    guidelines, 83
  on ozone controversy,
    55, 57
Weekly World News, 76
Westinghouse (firm), 146
White, William Allen, 121
Wilson, Edward O., 27–31
Wilson, Kenneth, 132–133,
  138
Wilson Research
  Foundation, 46
Wilson, Robert A., 46, 47
Wolfe, Sidney, 60
Women
  estrogen replacement
    therapy for, 45–47
  as Nobel prize winners,
    18–20
  in sociobiology, 27–30,
    141–142
Women's magazines,
  18–19, 45
Workers
  genetic screening of,
    123–124
  occupational health of,
    103–104
World War I, 87

XYY chromosome
  controversy, 84

Yalow, Rosalyn, 19

ABOUT THE AUTHOR

Dorothy Nelkin is a professor in the Cornell University Program on Science, Technology and Society and the Department of Sociology. Her research focuses on controversial areas of science, technology, and medicine as a way to understand their social and political implications and the process of decision making in complex technical areas. She is on the Board of Directors of the AAAS, Medicine in the Public Interest, and the Council for the Advancement of Science Writing. She is also a fellow of the Hastings Center and the AAAS and was a Guggenheim Fellow in 1984. She has served as an advisor and consultant to several organizations, including the National Academy of Sciences, the Office of Technology Assessment, and the National Council on Health Care Technology. Her books include *Controversy: The Politics of Technical Decisions, Science as Intellectual Property, The Creation Controversy,* and *Workers at Risk.*